THE CWMCARN 1

Tony Jukes

Danygraig Books, Machen
2011

Published in 2011 by Danygraig Books, 26 Danygraig, Machen, Caerphilly, CF83 8RF.

ISBN: 978-0-9571846-0-2

CONTENTS

Preface and Acknowledgements

I joined the Industrial Archaeology Society at Oxford House Adult Education Centre in Risca in 1971 soon after arriving in South Wales to work as a research chemist at the Waterloo Works, otherwise known as the *Paint Works, in* Machen. Oxford House Industrial History Society, the society's present name, has always been a vibrant society whose members have a great enthusiasm for the subject. The society meets weekly and runs its *Risca Industrial History Museum* in the former Colliers' Institute at Risca. I must thank society members past and present for developing my interests, research and increasing knowledge of the transport infrastructure and industries of the Rhymney, Sirhowy and Ebbw Valleys.

The catastrophe at Cwmcarn was one of these studies. The Ebbw Valley was the scene of many disasters in the nineteenth century. There were many explosions at the Risca Black Vein Colliery before that in 1860 which killed 142 men and boys. The valley was rocked by an explosion at the Abercarn Colliery in 1878, in which 268 men and boys died, and another at the North Risca Colliery in 1880, which killed 120 men and boys. Before these last two explosions a now almost forgotten disaster occurred at Cwmcarn in July 1875. A day of extremely heavy rain resulted in considerable flooding in South Wales. The earth dam of the Monmouthshire Canal Company's reservoir on the Crumlin line of their canal was washed away and twelve people were swept to their deaths. Then little used and despite the known dangers associated with such dams, maintenance had been neglected. An earth dam had been washed away at Holmfirth in 1852 and another near Sheffield in 1859 with much loss of life. No subsequent legislation was brought in to regulate the construction and use of such dams, as Her Majesty's Inspectors of Mines overlooked that industry, and many continued in use. No one was blamed for the failure of the dam at Cwmcarn. No official investigation was carried out to determine what happened and the verdict of the jury was inconclusive. Neither the Monmouthshire Canal Company or its officers and engineers were questioned or censored. When an application was made for the cost of rebuilding the flannel factory which had been destroyed by the flood water released by the destruction of the dam, the company's minutes state: *"That Mr H. Brain be informed that the funds of the Company are not available for the purpose of assisting him to rebuild the flannel factory which was destroyed by flood water from Cwm Carn Reservoir."* No compensation was paid.

In particular I would like to thank Mr Keith Harries of Bassaleg but formerly of Cwmcarn for furthering my research into the Cwmcarn dam disaster. Keith Harries grew up in Cwmcarn and became an electrical engineer underground in Cwmcarn Colliery. In 1990 began to take a great interest in the remains of this dam, particularly to safeguard its future in the event of any local development plan by Islwyn Urban District Council. With Mr Owen Gibbs, a retired civil engineer, he took some measurements of the dam and made the local council fully aware of the site's significance and history. He wondered if the area's mining history could explain the slump in the centre of the embankment. There was no mining activity in the area when the dam was constructed, but in 1837 the Number 6 shaft of the Monmouthshire Iron and Coal Company was sunk to the Tillery seam at the end of the parish road, now Nant Carn Road, and continued to work for many years. Coal from this mine was taken by a tramroad to the canal, soon to cross it by a bridge to Hall's Tramroad at Pontywaun Bridge, which tramroad became North Street in Pontywaun. Beyond the Level Cottages was a coal level, marked as 'Old Level' on an Ordnance Survey map surveyed in 1873 and published in 1879. This level is known to older people in the area as 'Bob Usher's Level'. According to Mr Harries, Usher continued to work this level in the Mynyddislwyn seam until Mr Griffiths, an agent for the Llanover estate who had gone to live at Abercarn Fach, exerted some influence to prevent this area being undermined. There was also a problem with ingress of water. This coal seam was not deep. Little is known about the working of this level. No consideration has been given to the possibility of mining subsidence affecting the dam, although the most likely cause is a slump in the structure due to leaks through its clay core washing away the materials of its construction.

Thanks are due to Tom Maloney and Phil Hughes of the Fourteen Locks Canal Centre, Newport Reference Library, and Foster Frowen for their assistance. Illustrations are from the author's collection and those of other OHIHS members who have made them available to *Risca Industrial History Museum*. I am extremely grateful to Malcolm Johnson for reading my script, his advice and encouraging me to finally publish my research.

Tony Jukes

Construction of the Canal and its Reservoirs

The Monmouthshire Canal Company obtained an Act of Parliament in June 1792 to construct a canal from the Town Pill in Newport to Pontnewynydd, with a branch from Crindau farm to Crumlin. Tramroads were to connect the canal basin at Pontnewynydd with the Blaendare and Blaenavon ironworks and that at Crumlin with the Beaufort, Ebbw Vale and Nantyglo ironworks. The Crumlin branch was to be 11 miles long and rise 358 feet by 32 locks: five at Alltyryn, fourteen at Cefn and seven at Abercarn.[1] On 10th July 1792 the Committee appointed Thomas Dadford Junior as their engineer, then working on the Leominster Canal. Work began immediately. At a General Meeting of Proprietors of the Company held on the 20th October 1794 in the Westgate Inn, Newport, Dadford reported on the state of the Crumlin line: *"The railroad from Beaufort to Crumlin, to the head of the canal, a length of 10 miles, with a branch to Sirhowy of 2½ miles is complete. Upon this road are two large bridges built over the River Ebbw Fawr and Ebbw Fach, with several smaller ones and culverts over the brooks. There are several branches of rail road made by other companies to join this road from limestone quarries, mines and ironworks. At Crumlin a weighing machine and wharfinger's house are built and weirs across the River Ebbw Vale for turning water into the canal. The canal from Crumlin to opposite the junction of the Sirhowy and Ebbw rivers below Carne Mill, 4½ miles, is navigable. This length is cut along very steep and rocky hillsides, interfered much with works and buildings and has been very expensive to make watertight. On it are 12 locks of 10 feet each, two large Aqueducts 30 feet high with embankments to them, eight road bridges, with a number of culverts, weirs, drain troughs and feeders for bringing water to the Canal. From the above place to Risca Lime Rocks, 1½ miles, is finished except a length of about 500 yards which wants bottoming and will be done in a month or six weeks. In this length are two road bridges, two deep embankments with culverts, drained troughs etc. From Risca Lime Rocks to the wood below Mr. Phillips, 1   miles, is mostly finished except about one third of the banking and the bridge at his house; and thence are other parts now in hand, equal to about 600 yards of level cutting. On this length is a great deal of rocky cutting, two bridges with two large embankments, culverts, drain troughs etc. From this place to the Cefn Public House, 1   miles, is let out and underhand, the puddle gutters mostly cut and parts finished equal to 200 yards .... From Cefn Public House to Alltyryn Farm opposite Maindee, 1½ miles, is not yet begun .... "* A meeting of the board on 5th December decided that all men working on the remainder of this line from Cefn to the junction, which included the '14 Locks', were to be removed to the Pontypool line of canal until that was completed. In February 1796 the main line of the Monmouthshire Canal was navigable from Pontnewynydd to the Town Pill in Newport. Construction of the flight of fourteen locks at Rogerstone delayed the opening of the Crumlin line through to Newport until 1799.

Its Act of Parliament gave the canal company power to abstract water from the Ebbw River and any stream within 2000 yards of the canal, with certain restrictions.[2] As Samuel Glover's Abercarn iron works had prior abstraction rights from Nant Celynen, Nant Gwyddon and the Ebbw River, and his corn mill at Newbridge was entirely supplied by the Ebbw River, the canal company were only allowed to divert water from those streams into the canal between the hours of 11 p.m. on Saturday and 11 p.m. on Sunday unless their ponds were full or water was running over the waste weirs belonging to the ironworks or mill. Furthermore, if the canal company built any reservoir above the ironworks or corn mill, or took any water from the Ebbw River above them, the Company were to erect across the Ebbw River at Crumlin all necessary weirs, flood gates and gauges as agreed with Samuel Glover for ascertaining the quantity of water in the river and regulating the quantity taken out by the canal company, taking no more than had been released from such reservoirs into the river. A waste weir was to be made on the side of the canal between the Abercarn upper weir and Nant Gowna for the purpose of discharging surplus water back into the Ebbw River.

To supply the Crumlin arm of the canal with water the canal company constructed the Pen-y-Van, Cwmcarn and Hafodyrynys reservoirs and abstracted some water from the Ebbw River. Water from the Pen-y-Van reservoir flowed down the Kendon brook, arrangements being made with a mill over the use of the water, and was carried across the Ebbw River to the canal by an iron trough supported on cast iron pier, later on a brick column, situated just north of the old road bridge at Crumlin. Water from the Hafodyrynys reservoir flowed down the Llawnan brook into the canal, the surplus passing into the river by the weir immediately below the old road bridge at Crumlin. Although now dry, the site of this reservoir can still be seen at Pont Pren and is occupied by the playing field of Hafodyrynys Rugby Club. The Cwmcarn reservoir was constructed in the valley of the Carn, which joins the Ebbw River on its east side and is surrounded on all sides by steep hillsides, rising to Twmbarlwm, the highest point in the area. The canal was supplied initially with water from the Ebbw River. At a meeting of the Committee held on the 23rd June 1795 it was decided: *"That Mr. Dadford do with all convenient speed proceed to repair the weirs across the Rivers Ebbw and Avon and to make such weirs as are necessary to bring the different feeders into the canal on both Lines."*[3] On the 22nd September 1795 the Committee ordered: *"That Mr. Dadford do immediately proceed to cut the feeder to convey the water from Carn Brook into the Canal according to the Plan produced this day."*

At the General Assembly of the Monmouthshire Canal Company held on the 25th April 1796 Dadford reported that the canal and rail roads to Cefn were complete, *"as they were last October"*, as was the line of the canal from there to Newport with the exception of one and a half miles of cutting and the fourteen locks down to Alltyryn.[4] Most of the lock pits had been dug out and a

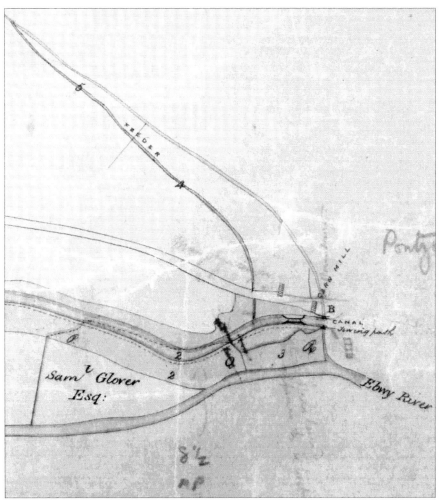

Mathew Williams' plan of the canal in 1795. The leat from Nant Carn which once ran to the corn mill has become the canal feeder and joins the canal above Cwmcarn Lock. The mill lies sandwiched between the turnpike road and the aqueduct. (Courtesy of the Fourteeen Locks Canal Centre).

start had been made on the construction of three locks and the cutting. A large quantity of good stone had been provided and arrangements made for the carriage of lime and sand to the site. Two or three reservoirs would be needed to supply the Crumlin line with water, on which there were three company boats. On the 28th June 1796 the Committee ordered: *"That Mr. Dadford do look out for proper situations for making reservoirs on the Crumlin Line of Canal, make Plans and Estimates of the same and report it to the next meeting."* Having constructed a feeder from the Carn Brook to the canal above the Cwmcarn lock, as shown on Mathew Williams's plan of 1795, the Committee on the 8th March 1798 ordered Dadford to construct a weir across the Carn Brook. On the 5th July 1798 the Committee ordered: *"That Mr. Dadford do immediately proceed to mark out the Ground necessary, and the necessary Trespass, for making the intended Reservoir on Carn Brook, and that notice be given to the Proprietors of Lands, of the Company's intention to make such Reservoir, and that they are ready to Treat with them for the same"*, and, *"That as soon as Mr. Dadford [h]as mark'd out the intended Reservoir on Carn Brook, Mr. Parry applies to Mr. H. Phillips, to value the Lands."* Four weeks later Dadford was ordered to begin construction of that part of the reservoir as far as the extent of Henry Evans' lands. The reservoir was formed by throwing an embankment, about 420 feet long and about 40 feet high at its centre, across the Carn valley. It had a surface area of nearly seven acres and was about 400 yards long, extending from Abercarn Vach as far as Abercarn Farm, almost up to the confluence of the Nant Gappy with the Nant Carn. On the 3rd September 1800 it was ordered: *"That a Sub-Committee of three or more of the Committee be requested to take views of the different Feeders and Situations for Reservoirs on the Crumlin Line attended by Mr. Dadford who is to report upon the same previous to the Committee being called to take the same into consideration."*

Leakage from the canal and its reservoirs was a problem from the earliest days. Mathew Williams's map of the canal in 1795 shows that there had been a breach on that part above St. Mary's church in Risca. Not surprisingly, in November 1795 the Committee ordered: *"That every boatman passing a stop gate between Carn Mill and Cefn should stop and put down the planks at each gate on pain of being discharged from working on the canal."* This indicates that there was a considerable danger of a breach in the canal caused by leaks, and that no stop gates* had been constructed on the long pound from Cwmcarn to Cefn. At a meeting of the Committee held on the 28th August 1799 it was

---

* Similar to a lock, to shut off the canal for maintenance on long lengths [pounds] between locks should there be a leak. Grooves are left in the masonry sides of bridge narrowings to insert planks for the same purpose.

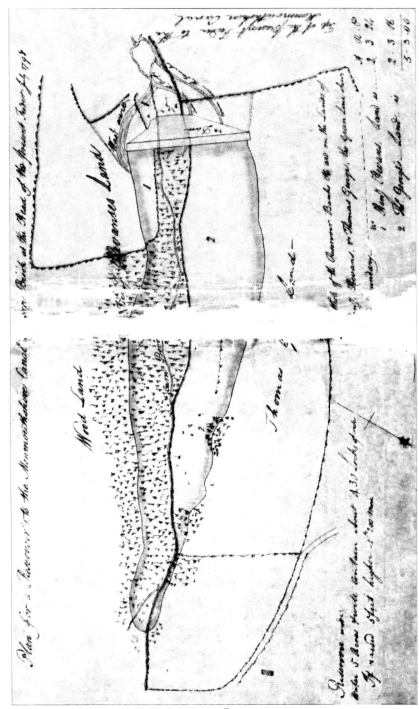

Mathew Williams' plan of the proposed Cwm Carn Reservoir in July 1798, to be built above the feeder take off on the brook.

reported that: *"there was a chasm across the road leading to Wayn Vawr* [Waunfawr] *coalery occasioned by canal water"*, and on 22ⁿᵈ September 1799, it was ordered: *"That Mr. Bage does on Saturday next let out the Ponds* [pounds] *on the Crumlin Line in order to stop all the Leakes he possibly can in that Day and the Water be let in on Sunday."* Bage was the canal company's surveyor. Presumably for reasons of safety and to prevent a flood following a major leak, he was instructed on the 12ᵗʰ February 1800 to put two stop gates in the Risca pound as soon as it was convenient to drain the canal. On the 10ᵗʰ August 1801 the Committee ordered: *"That Mr. Bage as soon as Harvest is over and can get Men do putt* [sic] *the Long Pond in Risca in proper repair, and not leave off untill* [sic] *the same is completely finished, and that Mr. Bage is desired to attend the men during the whole Time they are executing the same, and that due Notice be given the Freighters thereof."*

Edward Thomas Jones, lessee of the coal measures under Waunfawr from the Tredegar Estate, sent an account of the coal he had raised to the estate's agent Evan Phillips on 13ᵗʰ March 1800 and complained: *"I wish I had good cause to be pleased in the making out this Accᵗ. . . I have laboured in my mind Night and Day, I have spared no expense, I have given it a daily personal attendance. But I have had to contend with the varied seasons of the year, with misinformation, with an obstinate Man as to my Rail Road, with unfinished wharfs at Newport, and with a bad Constructed, ill managed, unfinished Canal; on which I had placed my every dependence, and on the Navigability of which I had made my first Calculations . . and which is confirmed by the sentiments or report of Mr. Outram, who declared the Canal to be in an imperfect state, and says that he is confident that the improvements he recommends will be found absolutely necessary before an extensive Trade can be established. On an extensive Trade alone, I have calculated; to be enabled to pay the rent, and produce a profit. . . Through the want of Water, Frost, Repairs, and alterations, the Crumlin line of the Canal was not Navigable last year above 8 Months."* [5] In July 1801 Jones again wrote to Evan Phillips, saying: *"The canal is a bad business and till some alteration by rail roads or otherwise take place I expect no profit from Waunfawr."* [6] There was a breach in the canal at Risca early in 1805, and William Davis presented the Company with a bill for injury and trespass on his lands occasioned by this.

On the 18ᵗʰ October 1802 Bage was ordered to inspect the puddled clay* on the embankment of the Carn Mill Reservoir, and take any steps to secure it

---

* Puddled clay is impermeable clay that is mixed with water to make it plastic and trodden in by gangs of workmen to give a waterproof core in earth dams and a waterproof bed along the length of the canal.

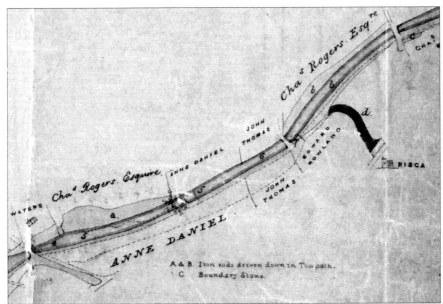

The Crumlin arm of the canal at Risca from Mathew Williams' map of 1795. d marks a breach in the canal bank above St. Mary's church. Iron rods have been driven down into the towpath at A and B, perhaps for bank stabilisation. (Courtesy of the Fourteen Locks Canal Centre)

immediately if necessary. On the 16th January 1805 the Committee ordered: *"That a Man or two be employed upon the Carn Mill Reservoir to stop the leakage from the same."* On 14th October 1814 the Committee resolved that Mr. Cooke should look at the Carn reservoir with a view to its improvement for the purpose of holding more water. After consideration of his report at the next Committee meeting it was decided to put his suggested improvements in execution under his direction. In a letter to Cooke dated 22nd May 1823 a contractor said: *"I beg to inform you of that we have opened the bank at Carn Mill Reservoir about 6 feet below the bottom of the Culvert but the foundation is very bad. I could wish to see you tomorrow morning if convenient before we begin to wall about the Culvert. We shall be ready for the masons about 11 o'clock. We should have been ready today but the weather has been very much against us."*

The minutes of the canal company reported few inspections of the canal and its reservoirs. On 20th September 1839 was written: *"The Committee this day inspected the Carn Reservoir together with portions of the Crumlin Canal and proceeded down the Crumlin and Sirhowy Tram Roads to the Canal House at Newport where they drew up a Report of their survey."* The Sirhowy

9

Tramroad joined the Monmouthshire Canal Company's tramroad to open a continuous line of tramroad from Tredegar to Newport's river wharves in 1805. The Monmouthshire Canal Company was permitted by its Act of Parliament to construct tramroads eight miles in length from any point on its line of canal without seeking any further power. Allowing for one mile of private tramroad constructed through Tredegar Park by Sir Charles Morgan, known locally as 'The Golden Mile' for its considerable financial benefit to him, the two tramroads met at the 'Nine Mile Point'. The Monmouthshire Canal Company eventually built a tramroad up the Ebbw valley in 1828, from the east end of the Risca 'Long Bridge' to meet that from the ironworks to Crumlin canal basin, to make direct tramroad connections between the ironworks and Newport.

Cwmcarn and the flannel factory viewed through Hall's Bridge, about 1905. Some of the arches were reinforced due to mining subsidence. The tramroad was leased by the GWR in 1879, who replaced the viaduct with a steel girder bridge which still stands today, although disused.

Benjamin Hall acquired the Abercarn Estate in 1808 on his marriage to Charlotte Crawshay, daughter of Richard Crawshay of Cyfarthfa, and began to construct tramroads from his mines to the canal. By 1814 his tramroad connected coal mines on his estates in the neighbourhoods of Manmoel and Gwrhay in the Sirhowy Valley and Cwmdows in the Ebbw Valley to his Abergwyddon canal basin at Abercarn.[7] Later, with the agreement of the Monmouthshire Canal Company, he extended this tramroad to meet theirs about one mile north of the Risca Long Bridge. This extension crossed the Ebbw Valley at Pontywaun, the existing replacement viaduct still being known as 'Hall's Bridge'. Thus, by 1830 the majority of the output of the ironworks and collieries in the Ebbw Valley to Newport had abandoned the Crumlin line of the canal and was carried on the tramroad. It is most likely that the Carn Reservoir was rarely needed after this date and suffered from neglect.

At a meeting of the Committee of the Canal Company on 20th January 1845 it was agreed that the Company would proceed with its Bill to make a railway from Newport to Pontypool, and take the title of the *Monmouthshire Railway and Canal Company*.[8] They would seek power to close the Crumlin line of the canal, and also the Pontypool line after making arrangements with the Brecon and Abergavenny Canal Company. There was great opposition to closure of the canal and this plan was dropped.

A watercolour by Lady Bunsen dated 27th October 1838 of the area which became Cwmcarn. The parapet wall of the aqueduct and the towing path over Nant Carn can be seen centre, the lock and lock keeper's house beyond it, and the Rhyswg in the background. (Courtesy of Newport Reference Library)

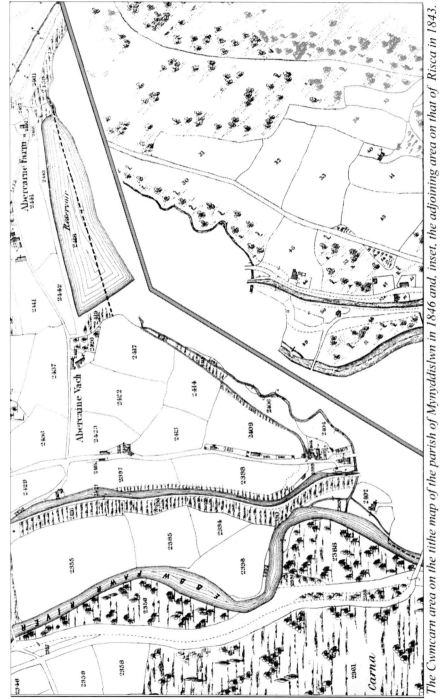

The Cwmcarn area on the tithe map of the parish of Mynyddislwn in 1846 and, inset, the adjoining area on that of Risca in 1843.

2

Cwmcarn

At Cwmcarn there was a corn mill and a flannel mill on Nant Carn, the latter close to the banks of the River Ebbw. In many places these shared the same water source. However, later Ordnance Survey maps do show a weir on the Ebbw River with a leat to the flannel mill. Cwmcarn flannel mill made hand woven cloth and used water power only to operate the fulling stocks.[*] The canal company intended to purchase the corn mill, demolish it and build a new one. The minutes of a Committee meeting held on 23rd June 1795 recorded: *"That William Griffiths of Lantarnam Millwright be applyed to for a plan and Estimate for Erecting a Mill at Carn Mill in lieu of the one to be taken down by the Company the same to be delivered at the next Committee."* A month later it was decided: *"That Mr. Brown and Mr. Henry Phillips be requested to meet with Joseph Phillips for Carn Mill, and to make the best terms they can with him."* The minutes of a meeting held on the 20th October recorded: *"That Mr. Butler and Mr. Nicholl be requested to give Instructions to Mr. James Jenkins of Caerleon to prepare a conveyance of Carn Mill and premises thereto adjoining from Joseph Phillips to the Company."*

On Mathew Williams's map of 1795 showing land taken for the canal Carn Mill lay near Nant Carn between the canal aqueduct and the turnpike road bridge, although any mill pond was probably destroyed by the construction of the canal. The feeder, either a new construction or re-use of the mill leat, ran into the canal above the lock. On the 8th March 1798 it was ordered *"That Carn mill be immediately altered to a Lockkeeper's house."* On 21st October 1799 the Committee ordered *"That Mr. Parry do get some proper person to value the Mill Stones at Carn Mill in order to dispose of them to the best Advantage,* and on 18th October 1802: *"That the Old Mill Timber at Carn Mill be sold Walter Walters at a fair valuation."* The old mill building and its situation was not a suitable lock keeper's cottage and so was decided on 2nd May 1821: *"That a suitable Lock Keeper's house be erected on the spot of Ground near the Lime House at Carn Mill belonging to the Company and that a person competent to sign permits and make a weekly return of all the bason trade be appointed to occupy the same."* Mary Phillips of Chapel Farm wrote to the Company on 3rd April to complain that the lock keeper had possession of a

---

[*] A fulling mill [a pandy] cleaned and thickened finished cloth by soaking the cloth in water with a detergent such as Fuller's Earth and beating it with the stocks. These were driven by water power. It caused the wet wool fibres to mat together and thicken the cloth, and removed the natural oils from the wool.

piece of her ground which he was using as a garden, and how the mason's came to build the lime house, now converted to house the lock keeper, she did not know.

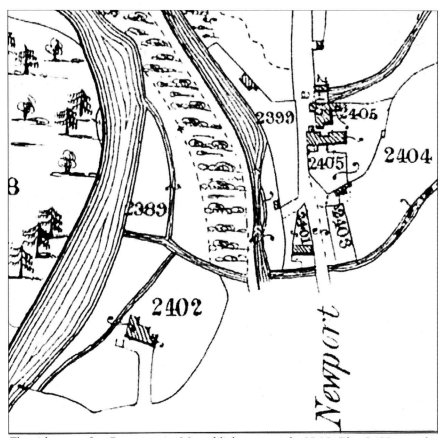

The tithe map for Cwmcarn in Mynyddislwyn parish, 1846. Plot 2402 was the lower woollen mill. The upper building in plot 2405 was the upper woollen mill which used the canal feeder for water power. Plot 2401 was the former Carn Mill, then just a small cottage between the canal aqueduct and the road bridge.

The corn mill was rebuilt on Nant Carn above the reservoir, near to the site later selected for Cwmcarn colliery. Little remained of the building but the millstones in 1970, but even these were removed a few years later during the construction of the *'Forest Drive'*. It would appear that Joseph Phillips had not met with the Company to agree terms as on 3rd May 1803 the Committee ordered: *"That an Order respecting Joseph Phillips be continued and in default of appearance at the next Meeting the Water Mill be taken from him."*

On 30th November 1802 Joseph Phillips applied to the Committee for the use of the water running from the reservoir to the canal for the purpose of turning machinery at his smith's shop, to which they agreed. He was to pay one guinea per year subject to receiving a day's notice of stoppage. This appears to be the date that the feeder was diverted into the canal at the foot of Cwmcarn lock, thereby giving a head of water to power a water wheel and machinery. There were problems with this supply as the Committee meeting on 6th April 1803 requested his attendance at their next meeting to answer a charge made against the tenant John Jones. By 1826 it was decided that the water flowing to the [flannel] factory should be let to Thomas Evans from 1st May as tenant from year to year at the rent of £2 10s. per annum, and payment of £14 14s. for the arrears of rent due that day.

Also, in 1802 Mr Phillips of Risca applied to the Committee for the waste water weir near Carn Mill to be removed lower down the canal, so that the water could be used for his grist mill at Pontymister. It was resolved that he would be accommodated, although the supply would be subject to a month's notice, at a cost of four guineas per year towards cost of its removal and re-erection, to which he agreed. As the waste water weir still exists at Pontywaun it appears this decision was not acted upon. The watercolours by Frances Baroness Bunsen, sister of Lady Llanover, made in 1838 depict a rural Cwmcarn. In 1841 only 20 houses were listed in the census for that part of Mynyddislwyn parish north of Nant Carn, excluding Chapel Row. There were three farmers, a miller, and two wool manufacturers. Tradesmen included four flannel weavers, a tinker, cooper, sinker, collier and four masons. At the corn mill lived Edward Matthews and his wife, their five children and his elderly father. The lower flannel factory was in the hands of William Williams, 50, who had two resident weavers. The upper flannel factory was held by Thomas Davies who also had two resident weavers, one only 15 years of age.

The *Monmouthshire Iron and Coal Company* sunk their first shaft at Cwmcarn in 1836 as part of plans to build an ironworks, the Victoria Ironworks, and develop coal mines. This was the Abercarn No. 6, a shallow shaft just 60 yards deep down to the Rock (Tillery) seam. They also built twelve houses, Millbrook Terrace or *The Ramping*, which were demolished in the 1960s.[9] In 1838 Roger Hopkins and Sons of Plymouth, civil engineers, trustees in possession, gave notice to the Monmouthshire Canal Company at their AGM on the 3rd May 1837 that they wished it to construct a tramroad under the 'Eight Mile Clause' in the Company's Act from the canal to this colliery. This application was refused and the colliery company built their own tramroad, its route becoming North Road in Pontywaun. On 3rd August 1838 they applied to the

Company to put a draw bridge across the canal at the nine mile post for taking coal from their colliery in Cwmcarn to Benjamin Hall's tramroad.

Millbrook Terrace, Cwmcarn, otherwise known as Ramping Row, about 1910. On the right is the tramroad from the No. 6 colliery to the canal whose route became North Road, Pontywaun. Construction work has begun on the GWR line to Cwmcarn Colliery, which crossed on an embankment in front of the row.

The Monmouthshire Iron and Coal Company was bankrupt by 1844 and the Abercarn colliery was taken over by Messrs. Allfrey, in business as brewers at the Castle Brewery in Newport, who traded as the *Abercarn and Gwythen Company*. In February 1845 they chose Ebenezer Rogers to be their manager and it was probably Rogers who built Abercarn Vach. After Rogers' death in 1863 his widow and family continued to live there. Abercarn Vach lay on the north side of the reservoir, almost level with the dam. Mrs Rogers rented the ground below the embankment of the reservoir and fenced it off to keep her garden private, hence the reservoir was known locally as 'Rogers' Pond'. This house was demolished in 1952.

Cwmcarn Colliery began as a downcast shaft for the neighbouring Prince of Wales colliery at Abercarn. The Ebbw Vale Company sank their first shaft here in 1876, but did not operate it as a separate colliery until 1912. The second upcast shaft was sunk in 1914. Both shafts were 281 yards deep. The stream was the boundary between the parishes of Risca and Mynyddislwyn

Abercarn Fach, the home of Ebenezer Rogers, manager of the Abercarn Colliery until his death in 1863.

and between the later Urban District Councils. The canal feeder ran down the north side of the Carn valley, a deep valley full of trees and underwood. Feeder Row was built along and above the feeder in the early 1870s. Following the explosion at the Prince of Wales Colliery in 1878 which killed 268 persons and caused the closure of the colliery for several years, the 1881 census recorded that most of these houses were vacant.

The upper flannel factory, which used the canal feeder for its water power, was in the hands of William Stradling from Caerphilly in 1851 and of his son Edward Stradling in 1871. It was this factory that gave the name *Factory Trip* to this section of the turnpike road down to Nant Carn, which it crossed by a bridge thirty feet high. Near the bridge on the south side of the stream in Risca parish was a pair of cottages, the Level Cottages, now one house, with a coal level adjacent running up the valley towards Abercarn Vach. On the north side of the stream above the road bridge was another cottage. Next to the road on the downstream side was the former corn mill, described in 1875 as a small old cottage whose roof was on a level with the arch of the bridge, and then the canal aqueduct. The cottage was almost sandwiched between the road bridge and the canal aqueduct.

The lock keeper's house above the lock at Cwmcarn, where Abraham Dugmore lived in the 1870s. On the hill behind is Twyn Carn House, built in the 1860s for a local farmer but demolished in 2009.

Cwmcarn lock with its lock keeper's house was situated above the aqueduct. Here lived the canal company's agent and lock keeper Abraham Dugmore. Twyn Carn House was constructed most probably in the early 1860s on the hill above the lock keeper's cottage by John Edmunds, whose family farmed in the area. This house was occupied by general practitioner Francis Davies in 1881, by David William James in 1891 (after surgeon Davies had moved to Abercarn Fach), and was purchased by George Nurse, managing director of the North Risca Colliery, about 1895, but recently demolished.

A draft of the purchase of a two quarter shares in the lease of the flannel factory has survived.[10] By an indenture of lease dated 1st June 1831 Benjamin Hall leased to Richard Humphrey that part of Pontywaun farm in both the parishes of Risca and Mynyddislwyn with the flannel manufactory and house thereon for 99 years from 25th March 1831 at a yearly rent of £5 5s. On 6th April 1839 David Jones, as sub-lessee with consent of Humphreys, assigned the lease to Edward Jenkins and his wife Amy. Amy Jenkins died and Edward Jenkins by his will dated 5th March 1857 left these leasehold premises to his four grandchildren, William Edward Jones, Mary Jones (wife of Newport coal merchant John Jones), Ann Jones and Amy Jones. His will was proved in the Consistory Court of Llandaff on 7th May 1857 by his executors Edmund Edmunds of Risca and John Davies. William Edward Jones sold his quarter share in remainder of the term of the lease of the house, factory and other premises to Hunt for £45 and that share of Mary Jones for £50. Hunt would have bought the other shares in the lease of the lower flannel buildings for a similar sum, valuing the remainder of that term of years of the buildings and machinery at about £190. Humphreys had both the upper and lower flannel mills.

The view from Hall's Bridge northwards towards Abercarn, by Lady Bunsen, 27th October 1838. Near the banks of the River Ebbw can be seen the lower flannel mill. (Courtesy of Newport Reference Library).

In 1851 John Hunt, 25, a native of Llandewi Brevi, Cardiganshire, was a flannel manufacturer in Pontnewynydd and employed 11 hands, one of whom was his nephew James Matthias, a bobbin weaver. John was named after his father who was also a flannel manufacturer. He lived at the Pontnewynydd flannel factory with his wife Elizabeth, 31, born in Llanstephan, Carmarthenshire. Hunt was declared bankrupt in 1852, with debts amounting to £162 11s. 9d. and assets of £150 6s. 9d. but rebuilt his business.[11]

On Elizabeth Hunt's death John Hunt married Mary Ann Player, about 41 years of age, at Trevethin parish church on 22nd September 1858.[12] She was the widow of John Player, who had been a coal miner and grocer in Abersychan. Giving up the Pontnewynydd factory where he had apparently carried on a large trade, Hunt took the Cwmcarn flannel factory in 1861, where later he built a new house. Later, it was said that Hunt was well known by all the Welsh clergy, and Mrs Hunt was well known in the district, having flannel clubs[*] in many places. The 1871 census records that John and Mary Hunt and sons John, a weaver, and James, and daughters Letitia and Elizabeth, were living at the lower flannel mill in Cwmcarn. Their youngest son James was born in the parish of Mynyddislwyn, probably at Cwmcarn, whilst his elder brother and sisters were born in Pontypool.

Hall's Bridge by Lady Bunsen, 27th October 1838. This view southwards from Cwmcarn down the Ebbw Valley towards Mynydd Machen also shows the Lower Flannel Mill. (Courtesy of Newport Reference Library)

There were no resident domestic servants or factory workers recorded in the 1871 census. In July 1875, the flannel mill was occupied by John Hunt, aged 50, and wife Mary Ann, 52, and his sons John, 22, and James, 10, and daughters Letitia Mary, 21, and Elizabeth Jane, 19, a servant Elizabeth Weeks, 17, a factory hand, Mary Jones, 15, and two apprentice weavers from the Caerleon Industrial School[†], James Foley and George Klein, 15, who had been

---

[*] Families made regular payments to their local wool manufacturer throughout the year to receive a supply of flannel for making new clothes in winter.

[†] This school was opened by the Board of Guardians of the Newport Union Workhouse in 1858 for the support of children who had lost their natural guardians.

with them two years. Both were resident at the Caerleon Industrial School in Mill Street in 1871, when their ages were given as eight and nine respectively and their birthplace as Newport, but were probably about 10 years of age. George's sister Fanny Klein, aged 8 years, was also there. Elizabeth Weeks was born in Aberdare, the daughter of David and Ann Weeks, a collier, and in 1871 was 12 years of age and living with her family in Dumfries Street, Ystradyfodwg.

The cottage between the canal and the turnpike road was occupied by Howell Davies, 60, a widower for over twenty years, his son John, a collier, 34, and his daughter Margaret Davies, 38, who kept house for them. The 1871 census reveals that he was born in Ystradfellte, Glamorganshire, and was then aged 59. His late wife was from Bedwas and his son and daughter were both born in Mynyddislwyn parish, possibly in Cwmcarn, and were 26 and 33 years of age respectively. In 1875 the upper of the Level Cottages was inhabited by William Bowen and his wife and young child, whilst in the lower lived a collier named Uriah Gover* with his wife and four children. In 1875 in the cottage on the north side of Nant Carn opposite the Level Cottages was said to live an old man named John Morgan with his wife, and a young lad. However, according to the 1871 census he was 48 and lived there with his housekeeper Margaret Ashman, 60, and a young grandson.

---

* He was wrongly named Govert in newspaper articles describing the disaster.

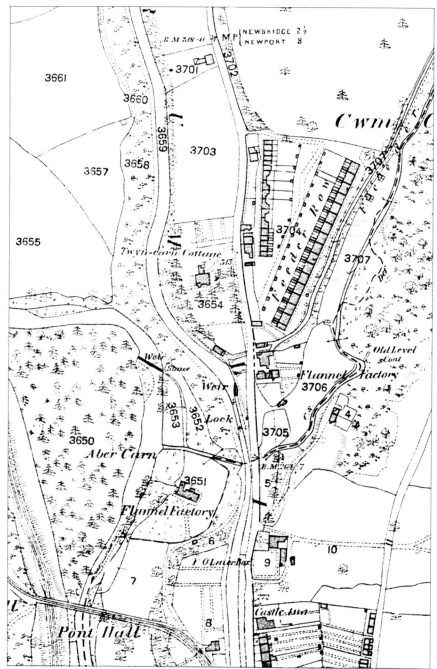

The Cwmcarn area as surveyed in 1873, from an Ordnance Survey map published in 1879. Only Feeder Row and Pond Row had been built

22

3

'And the rains descended'

Wednesday 14th July 1875 has long been remembered as the date of the great floods in Monmouthshire and South Wales. The barometer fell quickly as the rain came from the west. It began raining heavily early in the day and continued without intermission until early Thursday morning, when it gradually ceased. 5.3 inches of rain were recorded at Springfield, Monmouth, 5.31 inches at Tintern Abbey, 4.2 inches at Chepstow, and 2.71 inches at Bath. The rainfall decreased further east into England, with only 1.25 inches being recorded in Maidenhead and 1.29 inches in London.

The like of the flood had not been known for many years. Streams became raging torrents which carried away many bridges. Railways were torn up and traffic stopped, thousands of acres of land were flooded and crops, sheep and cattle swept away. Houses were inundated and hundreds of people driven from their homes. In Monmouth the Rivers Wye, Monnow and Trothy had flooded, and a further 12 feet of water were expected surging down later from Hereford. Rowing boats were the only means of access to properties on Monnow Street. Some wooden piles of the Redbrook railway bridge were washed away, with several railway navvies clinging to them. They floated down three or four miles, chased by watermen in rowing boats, before being rescued.

In Blaenavon a large sewer under the houses in Broad Street had fallen in two places, and the inhabitants of the row of houses were taken out of their beds at 2 a.m. through their windows. The stables of W. Gant, contractor, were completely washed away. Near Cwmavon, railway traffic was stopped when part of an embankment being the sidings of Messrs Vipond near the foot of their incline was washed away and nine trucks on their siding were deposited onto the Monmouthshire Company's line. Repair work was estimated to last a week, passengers having to walk about 400 yards past the obstruction between Blaenavon trains and Pontypool trains. Workmen at Abersychan dug trenches to carry away water overflowing a reservoir belonging to the British Ironworks, averting any danger. Brickyard Row, near the works, had to be evacuated and the inhabitants had great difficulty in rescuing their furniture. There was a landslip near Varteg colliery. Water washed at the foundations of houses alongside the stream in the Lower Cwm at Cwmynyscoy, where a Mr Painter lost part of the foundation of his house, his brewhouse and a privy.

At Pontnewynydd the road was flooded to a depth of three to four feet in front of the brewery and flannel factory. Too much water was flowing down the sloping ground behind the brewery for the small brook to cope with. A wall at the rear of Walters' Eastern Valley Brewery was completely swept away and his stock-in-trade destroyed. The flannel mills, where James Hunt had once been in business, were reported as destroyed. Houses opposite the factory had water three feet deep flowing through back doors and out of windows, panels of doors being broken to allow the passage of the flood.

The Avon Llwyd was a raging torrent. George Osmond's forge was flooded, as was his little grocery shop, and the machinery completely submerged.[13] The cascade opposite the mill just before Herbert's Wood was a miniature Niagara. The river tore down a large amount of wall on its south side, damaging the railway of the Pontypool Iron and Tin Plate Company, which company's pond also overflowed and caused a partial stoppage of the works. The Trosnant Brook also caused major flooding. In the dingle adjoining Albion Road a Mr Williams had to be carried from his cottage and the garden he was preparing for the Pontypool Flower Show was ruined. Several pigs were rescued before their sties were submerged. Cottages beneath the embankment fronting Mill Street were flooded and filled with mud. The gas works was flooded and the furnaces extinguished. Houses and the road below the gasworks at Trosnant were flooded with up to four feet of water.

In Pontypool, the rush of water in George Street above the bridge had ripped up the road surface and totally exposed the gas pipes. A cutting one foot deep had been made in the road near the monastery. On the Thursday morning the ponds in connection with the works were overflowing and the workmen opened the floodgates to prevent damage to the works. An immense volume of water was flowing over the weir near the mill, and a tree caught in the boiling surf below it was split as if by lightning. The flood in the feeder leading from the Town Forge pond to the waterwheel near the bridge at Park Road carried away part of the embankment. Mr Joshua's brewery and cellars were flooded as was his public house, *The Hanbury Arms*, which was flooded to a depth of four feet, the water sweeping the furniture into a heap in one of the rooms and forcing a wooden partition between two of the rooms. Mr Joshua and his family were rescued from their veranda. Inhabitants of several neighbouring houses were forced to leave their homes about 2 a.m., although an elderly Mrs Martha Banfield refused to go and yet survived the flood. A wooden bridge erected for Strick and Company's workmen near Pontypool Road was washed away. The old parish bridge to the Pontnewydd tin works was swept away about 9 p.m., as was a girder bridge on the Newport, Caerleon and Pontypool Railway at Ponthir.

The Crumlin road from Jarrett's Bridge to beyond Wain Wern and to the north of the road resembled a small lake. Water rushing through the bridge flooded into the bedrooms of the houses below, and swept away the footbridge giving access to Albion Road. More houses were flooded at the old Glyn Forge. Below Albion Road, gardens were covered by thousands of tons of tip waste brought down by the flood, and pigs were swept away. These families were forced to leave at 1 a.m.

At Upper Race, near Pontypool, houses had been built adjoining the Mountain Level at right angles across an insignificant stream that flowed from Cwm Lickey Mountain. The stream passed through a culvert and then through a drain under the floor of the middle house in the row. The culvert gave way and eight to nine feet of water flowed into the houses, carrying away doors, windows and furniture, thus gutting the houses. A stable and blacksmith's shop on the end of the row were destroyed. Escape was difficult, women and children being carried to safety on the backs of their men folk. The embankment was pushed down onto the Blaendare railway and the rails uprooted in three places, stopping traffic. It was reported that the Blaendare level, recently unwatered at great expense, had been flooded but this proved not to be true. Just below the Twmpath Level, five feet of water poured out of the New Mouth, although the workings escaped flooding, and thirty yards of an arch crossing the Lower Race Brook were swept away.

In the Ebbw Valley flood water nearly covered Crumlin station platform on the Monmouthshire Company's line. A large slip occurred on Mynyddislwyn Mountain opposite Abercarn, a new spring of water carrying earth down on to the Monmouthshire Company's line. This was deemed unsafe between Crosskeys and Abercarn, stopping the trains at Crosskeys station for several hours. One of these landslides was near the *Spiteful.*[*] A Monmouthshire Railway and Canal Company's engine had taken up a load of baulks to make the line safe. Returning with more wagons and a few vans loaded with workmen, another fall dragged 5 or 6 trucks and two vans loaded with labourers onto the road below, although none were hurt. Fortunately, the engine's coupling snapped and it remained on the rails. Several bridges were also said to be unsafe. Ballast under the private line of the London and South

---

[*] The *Spiteful* was a row of cottages erected by Benjamin Hall in 1810 down a hillside across the route of tramroad being built by Sir Henry Prothero from his collieries at Pen-y-Van and Manmoel to join the canal at the 8½ mile post at Cwmcarn. At that time Hall's tramroad terminated at the Abergwyddon canal basin. Although Prothero had not completed negotiations with Hall, he demolished some of the cottages to pass through.

The haulier for Edwards the grocer in Pontymister struggling in the flood water to rescue the horses from the stables.

Wales Coal Company was washed away and a great many trucks laden with coal were only just saved from toppling into the River Ebbw. A small pond at Jones's Darran brick works gave way and a slip occurred at the Rock Vein Colliery. The waters sprung up in a field belonging to David Morris of Graig House and carried tons of earth and rubbish down on to his farm house below, the occupants of which had to seek refuge upstairs. Risca was inundated and the roads impassable. A haulier employed by Mr Edwards, grocer, who had taken the horses to an adjoining stable, was carried away by the flood at Pontymister and nearly lost his life. Mr Edwards's shop and many other houses in Pontymister were flooded.

At Cwmcarn, a terrible disaster occurred when the canal company's dam gave way. The immense volume of water flowing down the valley carried away the turnpike road bridge and the canal aqueduct over the Nant Carn, a collier's cottage and the lower flannel factory, and flooded several other houses. Twelve persons died but three families had a lucky escape.

It was reported that Abraham Dugmore, the canal company's agent, had been at the reservoir at 2 p.m. as it was already full and opened a sluice. At 6 p.m. Thomas Thomas, gardener at Abercarn Vach saw three inches of water flowing over the embankment, which he did not consider unusual as he had seen the reservoir overflow the previous winter. George Stott, a gamekeeper who lived at Abercarn Farm at the head of the reservoir, said that a stream of water 70 yards wide was flowing over the embankment at 7.30 p.m. Thomas Evans, a sinker who lived at New Buildings at the top of Factory Trip, left the pit where he worked at 10 p.m. and tried to go home by way of the pond, arriving at the pond about twenty minutes later, but he could not cross as six to nine inches of water was coming over the embankment. He went down to the turnpike road and crossed the bridge, arriving home at 10.50 p.m. It had not struck any of them that there was any danger.

At the flannel mill that evening the apprentices had been sent to bed by John Hunt's daughters at 9.30 p.m., a little earlier than usual. They previously slept in the house but when the servant girls came they were moved into the old house, and then used as an extension to the factory. William Airey, a carpenter, had been visiting, said that all the family except his friend John, Hunt's son, retired to bed about 10 p.m. When he left to go to his home on Factory Trip a few hundred yards away, about 10.45 p.m., John suggested he should stop and lay down on the flannel as it was raining so heavily. However, he left the factory and ran through the gate by the old stable to the canal bank, crossed the lock, walking and running to his house. There was such a rush of water that he could not have used the footway under the aqueduct, which would have been his usual route. The canal was not overflowing at that time. He told his sister that he would take her down to see the Hunts the next day. He understood that Mr Hunt intended to go to Aberavon for a month's holiday on Thursday with his younger son James, who was in ill health.

Lewis Morris, a collier from Risca who had been living in Feeder Row for less than a month, had leant over the wall at the canal bridge at 8 p.m., looking at Hunt's house, and then walked down to Hall's Bridge where he also stood and watched the flood water, describing it as very muddy. When he returned to the canal bridge, he saw John Hunt and his wife digging clods and putting them against the fence to keep the water out of the garden, which was well-stocked with fruit trees and flowers. He was joined by John Woodruff, and expressed to him his concerns about the colour of the water, saying that it looked red and 'floody' and that it was not rain water or surface water but water mixed with earth. Morris said to Woodruff, *"It's a pity the people don't know about this water"*, and he replied, *"If you do tell them, perhaps they will not take much notice"*. Morris then said, *"I think the water must be oozing from some new breakage"*, but Woodruff replied, *"Tut, nothing of the kind"*, and they went home.

27

The breach in the earth dam of the reservoir (The Illustrated London News).

Morris was going to bed about 11 p.m. when he heard a great rending and cracking of trees and the rush of water. He lit the candle in his lantern and dressed. He went outside and found that he could not cross the feeder as usual owing to the flood water. He went back inside to tell the other couple who lived in his house that Rogers' pond had burst and then went down the row with his lantern as fast as he could. By the waterwheel at the upper factory the feeder had overflowed and after ten yards he was in water up to his knees before deciding to turn back. He decided to go in again and met John Morgan coming through the water with nothing on but his shirt. Morgan was confused. Morris went to take Morgan back to his house, but Morgan stopped at the first house in the row as he said he knew someone there. Morgan was trembling with fright and could hardly talk, and so Morris knocked the door until he aroused the occupants and heard them getting out of bed. He then carried on to William Airey's house on Factory Trip, where they were up and seemed in a bustle, calling out *"Rogers' pond has broke."* He carried on into water on the turnpike road and looked over to try to see Howell Davies' cottage, but it was covered by water.

Above the turnpike road bridge the water released by the breaching of the reservoir was impounded and flooded the Level Cottages to a depth of 4˙5 feet in the upstairs bedrooms. Bowen placed his wife and child on a high shelf of the cupboard, opened the window and swam around the back of the house with

The Level Cottages, once the home of the Bowen and Govert families and flooded to a depth of 4 feet 6 inches in the bedrooms, but now one house.

How the Govert family escaped from the flood water in their bedroom in the Level Cottages.

his child. Having left him safe, he returned for his wife. Govert, unable to escape through the window, stood his children on the window sill and used a piece of the iron bedstead to knock a hole through the brickwork in a recess and then led his wife and children to safety.

On the opposite side of the dingle was the cottage of an old man, his wife and a young boy. One account said he escaped from the house over the roof to get help to rescue the others, his wife being up to her chin in water when it suddenly subsided. This was John Morgan's house. Another account states that he was looking at the turbulent flood when he was thrown off his feet over the hedge into the garden, and his wife was left hanging out the window.

William Airey and his brother-in-law were joined by Lewis Morris, and together they tried to view the Hunt's house across the lock, but saw the side of the canal was gone. They saw a light in the house, and Airey shouted out, *"Thank God, those are saved"*, knowing that Howell Davies' house was submerged. William Airey had to be held back by the others, who told him he would lose his life if he attempted to cross. The water was rising and they got out as quick as they could. Just as they reached the turnpike road, they heard a terrible crash, and Lewis Morris said to William Airey: *"There it goes, there is a break somewhere"*, and the water level dropped. Also, their light went out. William Airey knew the canal had given way because the lock was empty. The water level was said to have reached the top of the aqueduct before it gave way. They went to Dugmore's house, the canal company's local agent, but his daughter said he was out.

The ruins of the Davies family's cottage between the remains of the aqueduct and the road bridge, with John Morgan's cottage above that. (The Graphic)

With the bridge and aqueduct destroyed there was only one way to get across the flooded Nant Carn to the Pontywaun side. They took the old road to Mynyddislwyn, crossing the River Ebbw at Chapel Bridge and the railway line and onto Hall's tramroad. They walked along the tramroad to Hall's Bridge and crossed back over the River Ebbw, from where they could see the white-painted factory but not the house, and still were unable to get across because of the flood water. Airey managed to get into the stable, up to his waist in water, as his friend had mentioned that he would take their pig there, but it was empty. They lost their light and went home until daybreak.

Water apparently began to overflow the canal bank below Trinity Chapel, Pontywaun, at the rear of a row of cottages about 10 p.m. and people had been to Dugmore's house to tell him and ask him to raise the sluices and let some water out of the canal. Dugmore went out and removed planks from the overflow and then tried to raise the sluices at Pontywaun. He was joined by two police constables who were out searching for him, whom he asked to cross over by the canal bridge and help him raise the sluices. Before the aqueduct gave way, water was overflowing the canal just above the Eagle Inn and rushing down the turnpike road towards Crosskeys station twelve to eighteen inches deep, flooding houses on the side and damaging carpets and household effects. With the water rising, they beat a hasty retreat to the canal bridge. The rain was falling heavily and it was very dark. They lost sight of Dugmore and feared that he had been washed away, but he was found a little further along the canal bank by James Holvey and others who had come out of their houses. Suddenly they heard a shout, *"Run for your lives"*, and hearing a terrific crashing noise and a rush of water, they ran back to the canal bridge.

There was intense anxiety in Risca and Pontymister about the canal bursting its banks. Water spouts were springing up on the mountain sides and water

People wading across the flooded road near the old Risca Long Bridge.

was pent up near the brickworks of Messrs Greene and Company. Water flowed down the road three or even four feet deep and did not begin to subside until Thursday night. P.C. Lloyd of Abercarn had been on duty that night, walking down to meet a colleague, Police Sergeant Williams of Risca, and had been on the Carn Bridge ten minutes before it collapsed. Sgt. Williams had to wade through water waist deep to reach his own house.

31

The ruins of John Hunt's house and flannel mill as seen from the remains of the aqueduct. Water still flows from the feeder below the bottom gate of Cwmcarn lock. (The Graphic)

There was a wide breach in the canal, about forty to fifty feet deep. Lights were obtained and Dugmore organised a rescue party, comprising James Holvey, William Holvey, Jacob Williams, Daniel Allsopp, George West, Mr Benjamin and P.C. Lloyd. They plunged into the flood water, holding hands for safety, and found James Foley clinging to the rafters of the flannel factory. He was later to say that he was woken by a crashing noise, the water coming in, and grabbed the warping frame which was fastened to the rafters above their bed. He had no chance to George Klein, the other apprentice boy. Drawn by the barking of a dog, they found Mr Hunt clinging to a small ash tree on the river bank. At first the dog would not let them near but he was eventually taken to a friend's house. The search for others was abandoned until dawn.

All local work was suspended the next morning to allow people to search for the victims and to assist those whose homes were ruined by the flood. William Bowen and Uriah Govert were outside their houses, furniture and bedding heaped up and covered with mud and sand and totally unfit for future use, telling everyone how they escaped and saved their families. Govert looked pale and ill. Very little remained of Howell Davies' house. It was believed that the turnpike road collapsed onto the submerged house when the canal gave way. A number of men turned over the debris of Howell Davies' house until darkness, convinced that they would find the bodies buried beneath it. The canal arch through which the water ran still stood in the midst of a gap at least thirty yards wide. At the flannel mill Mr Hunt's house was completely gone and the machinery of the flannel factory lay in a shapeless mass, the waterwheel embedded to its centre in mud and stones. A *Daily News* reporter said: *"How it came about that a house and factory were built just in the line that a flood would take if the pond burst, we shall probably never be told. A man may settle where he likes, and no one may call him to account."*

Nets had been fastened to the Cymmer Bridge[*] at Crosskeys to catch any bodies and scores of people walked along the river bank searching for the victims. Joseph Hawkins, a collier at the Rock Vein pit for nine years and living at the back of the Crosskeys Inn, found the body of Mrs Hunt by the Black Vein Colliery at 2 a.m., having seen something white moving about in the water. Her body was taken to the Crosskeys Hotel, the landlord being awoken to receive it.[†] By Thursday night, eight bodies had been recovered. Those of Elizabeth Hunt, James Hunt and Margaret Davies were found in the Pandy fields. The body of Letitia Hunt was found also found near the Black Vein pit. P.C. Seys found the body of George Klein near Pontymister. Mary Jones's body was found near the old Risca Copper Works and that of Elizabeth Weeks near Tydu, Bassaleg. The body of John Davies was finally found on Saturday in a field opposite the Black Vein pit, but that of his father, Howell Davies, was believed to have been carried out to sea.

Thomas Lewis, Baptist minister at Moriah Chapel, Risca, recorded in his diary: *"Very heavy rain and storm. The pond at Cwmcarn burst, and the flood rushed down from Cross Keys to Risca with terrific force. The canal and the turnpike were cut through, and the waters of the canal joined the flood. John Hunt's factory was destroyed, his house swept away, and the family of eleven perished. Another house near, inhabited by Howell Davies, his son and daughter, was also swept away, and all three perished. The body of Howell Davies was never found. Risca and Pontymister were flooded. Water in most of the houses, gardens spoiled, roads torn up. It will take hundreds of pounds to repair the immense damage. All this happened at midnight. .... The scene from the pond to the bridge below the factory was awful. The same night a waterspout tore up the field by the old copper works and caused much loss and damage."*[14]

Land near the flannel factory was covered by six feet of debris. Pieces of flannel, linen and clothing festooned the trees and bushes along the river bank. Colliers flocked to the area as they would to the scene of a pit disaster and helped rescue articles but also filled the public houses.[15] Furniture, clothing, bedding, books, bibles and papers were picked up and carried back to be heaped up around the house where John Hunt lay, but a great deal of clothes and rolls of flannel were taken away by the finders. There were heaps of

---

[*] The old bridge taking the road from Crosskeys to Pontllanfraith across the River Ebbw, near the present Pandy Park.
[†] It was customary for the bodies of accident victims to be taken to the nearest public house, and for the inquest to be held there.

The ruins of John Hunt's house and flannel factory. In the foreground can be seen the remains of the waterwheel. On the right, past the remaining abutment of the aqueduct, are the Level Cottages. (David Thomas Collection)

Rescuers search the ruins of the cottage that lay between the aqueduct and the road bridge to find the bodies of the Davies family. On the right are the Level Cottages from which the Bowen and Govert families had to escape from their bedrooms. Behind the ruins of the road bridge is John Morgan's cottage. Dugmore's house, the canal agent, can be seen just above the lock, with Twyncarn House behind it.   (Illustrated London News)

broken furniture and damaged goods about the village. Rumours said there had been a great deal of money in Mr Hunt's house, although a family friend said it was only about £20. The flannel club books were lost. The belongings recovered were taken away by Mrs Hunt's friends, and timber and other goods sold. Mr Hunt apparently had life insurance, and owned the factory and fields.

Friends help John Morgan salvage his furniture and belongings from his cottage, whilst others talk and view the turbulent Carn Brook. (The Graphic)

Hundreds of people visited the scene of the disaster the following Friday, the visitors stripping the tree to which John Hunt clung as souvenirs of the disaster. As he lay on his sickbed John Hunt was visited by scores of people, including many newspaper correspondents, and got little rest. A great number of friends arrived from Ystradgynlais to see him on Friday night. He was in great pain but never lost consciousness. William Airey visited John Hunt several times and spoke to him, and said that Hunt recognized his voice. Hunt was distressed at the loss of his family and frequently said he did not wish to live. He put his affairs in order, making his will leaving his property to relatives in Pontypool, and wished to die as soon as he could. He said there was a cash box in the mill containing £25, but this was not found.

Before his death on Sunday 18th July 1875 John Hunt gave an account of the destruction of the flannel mill.[16] Mr Hunt, his eldest son and William Airey had seen the water rising, but were not overly concerned. As he looked towards the pigs cote on the aqueduct side of the house he said to his son *"John, I think you had better go and fetch the pig out of the cote and put it in*

36

the stable" but John said *"Oh, we have seen the water as high as this many times"*, and did not do so. Mr Hunt left the factory and went into the house, which was divided from the factory by the old house, in which the apprentice lads slept, and went to bed, while his son lingered at the door talking to William Airey. His daughter Lettie called for her brother to make haste and come indoors. He got undressed and into bed then heard a loud noise. Looking out of the window he saw an immense volume of water flowing towards the house. He shouted to his wife *"Get up Mary, I am sure there is something the matter"* and called up the rest of the family. Mrs Hunt got up and slipped an old dress and a jacket over her nightdress. At the bottom of the stairs he saw his son John and shouted *"Run John, and open all the doors for a great flood is coming."* He did so, and was holding the front door when the water burst in and swept him out of the house, never to be seen again. Mr Hunt ran back up the stairs and gathered the family in an upper room where there was a ceiling hatch giving access to the attic. Boxes stored there filled with new flannel were used to climb up. He pushed his family through and climbed up himself. Looking to the gable wall he saw it crack. As he darted across to the front wall, which faced the viaduct, the front of the house fell with a mighty crash, and he was washed away. He was carried by the water into a small tree in Mary Kenvin's field, where he clung until he was rescued, although he could not have held on much longer. His dog had been washed into the same tree and swam to him, its barking drawing his rescuer's attention to him.

James Foley and George Klein, the two apprentices that Hunt had taken out of the Caerleon Industrial School, were in bed. Foley said he had heard and seen the water approaching as he was awake. He turned to George Klein with whom he shared a bed and said *"Get up George. There is a flood coming."* But George did not respond and did not get up. Meanwhile, stooping down and quickly gathering his clothes out of a box under his arm, he climbed on the warping machine which was fixed to a roof beam and then onto the beam and clung tight. He saw the front of the house collapse and the bed with George Klein in it was washed away He clung tight to the beam until he was rescued.

The reporter riding through the floods to the telegraph office with his story (The London Illustrated News).

4

The Funerals

On the following Sunday morning 18th July 1875 six of the bodies were buried and the bodies of the two servant girls were carried across the mountain towards Bedwas. Three graves had been dug in the burial ground at Trinity Chapel[*], Pontywaun, a double one for the five members of the Hunt family, one for Margaret and John Davies, and one for George Klein. Although the Hunt family did not attend Trinity Independent Chapel, the elders had given their permission for the family to be buried there. Their minister, Peter H. Davies, was a young man only recently appointed. The Rev. Thomas Lewis of Moriah Baptist Church, Risca, recorded in his diary that Hunt was a *"Freethinker"*. Thousands of spectators were present, some of whom had travelled a long distance, but excellent order prevailed. Undertaker Mr D. Tovey arranged the order of procession and the six coffins were carried from the Crosskeys Inn on the shoulders of bearers one behind the other to the burial ground. Police Superintendent McIntosh and Inspector Sheppard had stationed police officers there to keep order. The Rev. Peter H. Davies met the mourners at the gates and led them into the chapel, followed by the coffins. The chapel was full and the doors were closed. He read Psalm 90 and Corinthians Chapter 13 and gave a brief solemn address, hoping they would well understand his utter inability to say anything that would add to the impressiveness and solemnity of the sad circumstances they found themselves in. He told the congregation that life here at best was short and transient, they being subject to the howl and blast of the storm, he said, and unless they were fixed upon the Rock of Ages they would be lost in this life and for ever and ever. He asked if there were any ungodly persons in the chapel, pointing their attention to the graves and scenes outside, and urged them to accept the Divine offer of mercy before such a calamity should befall them. In the evening, he preached from Matthew, Chapter 7, Verse 27: *"And the rains descended, and the floods came, and the winds blew and beat upon the house, and it fell; and great was the fall of it."*

John Hunt died at 3 a.m. on Sunday morning 18th July, hours before the burial of his family. After his death a large rusty nail was found embedded in his hip. On Monday afternoon 19th July, his body was carried across the canal and conveyed in a hearse to the chapel and buried with his family. Strong feelings

---

[*] Trinity Chapel was built 13 years previously with the aid of a fund set up by Samuel Morley MP for the building of Independent Chapels in South Wales where the English language prevailed.

Trinity Chapel, Pontywaun. The Hunt family's memorial is visible on the right edge, centre. About half of the memorials are to victims of colliery explosions.

were expressed about the hasty way in which he was buried. Few people knew about it and none of his friends were present to mourn his passing. John Davies and Margaret Davies were also buried on Monday at Trinity Chapel. Again, only a few that knew about it attended their funeral. Their father Howell Davies was a Calvinistic Methodist and a *"godly man"* according to the diary of the Rev. Thomas Lewis of Moriah Baptist Church.

The memorial to the Hunt family, the inscription of which once read: "In loving memory of John Hunt of this parish, flannel manufacturer, aged 47 years and Mary Ann his wife, aged 52 years John aged 23 years Letitia Mary aged 21 years Elizabeth Jane aged 19 and James - aged 11 years their children, who with two servants and an apprentice boy were all swept away by the flood on the night of Wednesday 14 July 1875 by the bursting of Cwm Carn Reservoir. Mr Hunt survived only for a few days being found holding to a tree and died on the following Sunday from injuries received from the Catastrophe. "And in Death were not divided". Erected by his sorrowing relatives."

40

5

The Inquest

A *Western Mail* reporter writing about his visit to Cwmcarn in the *Castle Inn*, Pontywaun, on Friday 17th July 1875 following the disaster and the day before the inquest was due to open contemplated the cause of the disaster. Was it due to a latent defect in the reservoir and, if so, how far was it defective? Was the sluice capacious enough? Was the sluice in good working order and properly maintained? He hoped these questions would be answered by the Coroner's inquest. A Coroner's inquest is an inquiry into the circumstances surrounding a death to establish the identity of the person and how, when and where they died. It has to establish the details needed by the Registrar of Deaths to issue a death certificate (since 1874), but the inquest is not a trial and it is not the Coroner's role to find people blameworthy and apportion guilt. Although people can volunteer to give evidence the Coroner decides who to summons as witnesses.

The inquest was opened on Saturday morning 17th July by coroner William Brewer at the Crosskeys Inn. The bodies of the victims were laid out in the inn's clubroom where they had been taken as they were recovered by permission of landlord Ezra Davies. The jury were:

    John Williams, Pontymister, grocer (foreman)
    Thomas Williams, Risca, shoemaker
    Walter Rosser, Moriah Tump, builder
    Charles Harris, Pontymister Farm, farmer
    Vincent Allen, Risca, grocer
    Frank Masters, Risca, watchmaker
    William Howells, Risca, draper
    Rev J. W. Winspear, Risca, Wesleyan minister
    Alfred Summers, Risca
    Thomas Harris, Ochyrwyth, farmer
    William Griffiths, Pontywaun, draper
    James Silverthorn, Risca, shoemaker
    William Hartshorn, Pontywaun
    Jesse Silverthorn, Pontywaun

The jury viewed the bodies in the clubroom, all of whose faces were reported to bear disfiguring injuries caused by the floods, although another report said that the countenances of the dead were *"an aspect of peace and serenity"*. A number of people hanging around outside were allowed then to view the bodies, and were much moved by the sickening sight. The coroner and jury visited the scene of the disaster and the reservoir, to examine the embankment in order to ascertain the cause of it giving way. They told the coroner that an engineer should examine the dam and advise them in order to judge if the Monmouthshire Canal and Railway Company were in any way blameworthy. He agreed with their request, but said that the reservoir had stood the test for ninety years and this was the first time it had given way. After returning to the Crosskeys Inn, the inquest was adjourned until Monday 26th July 1875.

At the resumed inquest the Monmouthshire Railway and Canal Company were represented by their solicitor Mr H. Gustard Stafford, the family of Howell Davies by Mr C.H. Jones of Merthyr, and the Hunt family by Mr H.G. Lloyd of Newport. Joseph Hawkins, a Risca collier, was the first witness to give evidence. He had recovered the body of the first victim, Mary Hunt, and this was an inquiry into the cause of her death. Hawkins said he had found the body of Mrs Hunt near the confluence of the Ebbw and Sirhowy rivers, having seen something white rolling in the waves. He said the talk of the neighbourhood was that the reservoir was not safe, but had not been told it was dangerous or that it would burst and admitted he could not make any judgement. When fishing at the reservoir he had not seen any leakage or any workmen repairing the dam, although about six months previous the reservoir had been drained. The dam was built up of clay and stone with a stone batter, but he could not tell if it was puddled with clay. On Mrs Rogers' side there was a stone wall along the top. Under the embankment was a bed of watercress about three or four yards long and up to four feet wide, but not in a watercourse.

As watercress needs clean running water to grow the source of the water which supplied this cress bed needed to be established. Mary Ann Brown of Pontywaun said she had been to the pond frequently until her collier husband was put in an asylum ten months previous but had not heard that the dam was unsafe or seen any leaks. She knew the watercress bed, having eaten some of it, but would not cross the hedge as she dare not go into Mrs Rogers' garden, and thus did not know if the ground was spongy or boggy. She thought the bed was fed from a well some distance off. William Airey said that he had seen a heap of red clay on the embankment about four or five weeks before the disaster. Until the pond broke he believed there were only two sluices, but he now knew there were three. The central sluice was broken and useless but one of the other two could be used, although he knew nothing of that sluice enclosed in a box. Airey admitted that he was not qualified to judge the safety of embankment. Lewis Morris said that he had walked over the embankment and thought it was rather shallow for the weight of water behind it, perhaps ten or eleven feet at the top, compared with others that he had walked on and one that he had helped to construct.

William Cooke, a sinker, was called. He stated that water was not running over at 2 p.m. on the Wednesday 14th July 1875. He saw Dugmore there with his tools and he appeared to be going from one sluice to the other. Thomas David of Cottage Row, Cwmcarn, a collier who had been a keeper under Mrs Rogers for seven years, had walked over the embankment at 5.30 p.m. on the 14th July and had seen a hole on the top where the water later broke through. Water was flowing over the embankment two to three inches deep. He had seen Dugmore and men often repairing the embankment. He believed the deep sluice had not been worked for the last six or seven years and thought that the embankment was unsafe.

Thomas Thomas, gardener to Mrs Rogers since November, considered the dam was unsafe. In his opinion the roots of trees at the back of the bank were all that was keeping it together, otherwise it would have gone years ago. There was nothing but loose sand and gravel in the bank, which was lower in the centre. Water flowed through the embankment about five or six feet up the culvert that would need a four inch pipe to take the flow. Thomas said the watercress bed was fed by water from the culvert which fed the canal. Mr Rogers had made the watercress bed with the water coming from the culvert as the head of the cress bed was near the culvert. He did not know of a spring on their premises which fed it. There was also about two inches of mud on the bank about fifteen feet from the top of the embankment on the wood side due to leaks. He never considered the leaks dangerous. The deep water floodgate would not have let out all of the water coming in, a dozen flood gates would not have taken away that amount of water. Thousands of tons of earth and stones were fetched down from the hills. The wind had been blowing heavily and in the day the wind veered round and blew against the head of the pond. There was a dry stone wall on the embankment about three feet high. The water usually stood about one foot from the top of the embankment. Three inches of water was flowing over the embankment at 6 p.m. which was not unusual as he had seen it overflow the previous winter. He heard his dog barking outside at 2 a.m. and when he arose to let it in he was told of the accident.

George Stott, a gamekeeper who lived at Abercarn Farm at the head of the reservoir, said that a stream of water 70 yards wide was flowing over the embankment at 7.30 p.m., but he had not considered it dangerous. He had never seen so much water coming down the two dingles before, carrying with it trees, stones and rubbish. In his opinion, the three sluices would not have taken the water coming down. There were two leaks in the embankment, one place being particularly "*moist and gobby*". About two years ago, he had seen workmen repairing a hole about three feet deep in the centre of the path where the break subsequently occurred.

Thomas Evans, a sinker who lived at New Buildings at the top of Factory Trip, left the pit where he worked at 10 p.m. and tried to go home by way of the pond, arriving at the pond about twenty minutes later. He came down by the waste weir and tried to cross the feeder by the sluice, but the water was over the arch and the feeder. He then went back and tried to cross over where the water later broke through, but he again could not cross as six to nine inches of water was coming over the embankment. Also, a large tree had fallen across the embankment. He returned and went down to the turnpike road and across the bridge, arriving home at 10.50 p.m. It had not struck him that there was any danger, and he did not know if the centre of the embankment was lower than elsewhere. He lived about 100 yards from the bridge but heard nothing except the sound of wind and rain and went to bed at midnight. He arose at 12.40 a.m. as water was coming into the back of the house, but knew nothing of the disaster until the next morning.

Abraham Dugmore was called but on the advice of Mr Gustard representing the Monmouthshire Railway and Canal Company, declined to be examined. Even today, a witness called to give evidence at an inquest does not have to answer questions that may tend to incriminate them. Gustard was legally present to represent the interests of the canal company at the inquest and to question witnesses should any statement they made blame the company. The Coroner had failed to summons any officers, directors or employees of the canal company as witnesses other than Abraham Dugmore.

James Foley, now living at Merthyr, said George Klein woke him. He heard someone screaming, jumped out of bed, and then the house fell. He managed to grab the warping frame which was fixed to a beam, which saved him but he had no time to warn Klein. When he went to bed at 9.30 p.m. the water was coming into the garden. Neither Mr Hunt nor anyone else had ever told him the pond was unsafe.

William Airey said that he had never heard Mr Hunt say that the pond was unsafe, but his son had always it was so and possibly it would be the end of them. When he first came to Cwmcarn two years ago he had lodged with Dugmore, whose duties were to attend the lock and take tolls, and had seen Dugmore go backwards and forwards to the pond, especially on stormy nights. Airey had not seen any men repairing the pond, although they had cleaned the feeder. There was a cartload of red clay around the sluices about four or five weeks previously and he had noticed puddled clay around the sluices and blue clay in the bank.

The coroner summed up with great brevity, saying that floods of this magnitude were unknown in the district. He said the reservoir had been in existence for about eighty years, and asked the jury to consider if the embankment was safe and to record a verdict accordingly if neglect was to be attributed to any party. To record a criminal verdict, there must be evidence of gross and culpable neglect. The jury returned their verdict: *"That Mary Ann Hunt met her death on the 14th July, 1875, through the unsafe condition of the embankment of Cwm Carne pond, commonly known as Rogers' pond, the property of the Monmouthshire Railway and Canal Company."* The coroner then satisfied himself that the jury believed the company was guilty of neglect but were not criminally negligent.

The Coroner had failed to issue a summons to require the attendance of an engineer or surveyor working for the canal company with responsibility for the safety of the reservoir to give evidence. He did not call any director of the company, which might have revealed that the dam suffered from leaks from the time it was built. Although the jury asked that the dam should be examined by an engineer so that his report could be given at the inquest, this was not done. The result of the inquiry was inconclusive and the jury did not express any strong opinion despite hearing that:

(a) One of the three sluices was inoperable;
(b) There were two leaks in the dam near the subsequent breach;
(c) There appeared to have been little maintenance, apart from plugging leaks with puddled clay;
(d) The reservoir could not cope with the rainfall of the 14th July 1875.

From the testimony of witnesses Dugmore had done his duty that day as far as it was in his power. Clearly, there was enough evidence to ascribe the cause of the disaster to the culpable negligence of the Monmouthshire Railway and Canal Company as the dam was in an unsafe state. Without testimony from Dugmore or the company's surveyor, engineer and officers there was little more that could be said.

The minutes of the next Sub-Committee Meeting of the Monmouthshire Railway and Canal Company held at their offices in Newport on the 5th August 1875 record that: *"The Verdict at the Inquest held on the body of Mary Ann Hunt who was drowned by Water from Cwm Carn Reservoir on the night of July 14th was read and the exertions of Canal tender A. Dugmore in trying to prevent damage by flood water on that occasion having been reported it was Resolved - That he be awarded £5 in recognition of special services at great peril to himself."* [17] No other details or expressions of sympathy were recorded at this or any subsequent meeting of the Company.

6

An Investigation

The General Committee of the Monmouthshire Railway and Canal Company decided to ask Messrs. Hawkesley and Abernethy, Civil Engineers, to view and report on the cause and effect of the failure of the embankment of the Cwmcarn Reservoir, but the minutes do not record that this decision was acted upon or any report received. The Coroner's jury had asked for an engineer's report on the Cwmcarn dam but no investigation was carried out.

There had been two previous disasters caused by the failure of earth dams in Great Britain. A dam 98 feet high near Holmfirth gave way in 1852 and 81 people were killed. In 1859 near Sheffield, another 95 feet high gave way and resulted in the death of 250 people. Both floods caused immense damage.[*] These disasters prejudiced the minds of engineers against the earth dam and subsequent nineteenth century dams were massively built of masonry in accordance with a theory of construction propounded by Professor Rankine of Glasgow.[18] However, many earth dams remained in use without regulation and inspection, including that at Cwmcarn.

It was George John Jee, a mining engineer, 41 years of age and living in George Street, Pontypool, in 1871, who decided on his own account to conduct an investigation of the reservoir and its embankment. This he reported on at a meeting of the South Wales Institute of Engineers in 1877, and it was subsequently published in their journal.[19] The length of the reservoir was about 400 yards and its extreme width at the lower end was 419 feet. He calculated the area as 6½ acres, which broadly agreed with the accepted figure of 7 acres. The extreme depth of water opposite the centre of the embankment at its lower end was 35.5 feet, the measurement being taken from the top of the sill of the waste water weir. Without taking a series of cross-sections of the valley he was unable to calculate the mean depth as the bed was irregular. Taking the average depth as 10 feet he estimated the reservoir contained over 90,000 tons of water at the time of the accident. For an embankment 419 feet long and assuming an average depth of water of 24 feet (again without careful measurement) Jee estimated the pressure of water on the face of the embankment as 3,367 tons. This estimate did not allow for the strong easterly wind pressing the water from the head of the reservoir towards the lower end.

Jee's survey showed that the embankment had been built with a uniform gra-

---

[*] Details of these floods are to be found in the Appendix.

dient of 1 in 3 on its front or inner face and 1 in 2½ on its back or outer face. The embankment was 22 feet wide where the water broke through and 19 feet on the north side but getting wider from the centre to the extremities, which could not be an approved method of construction. The centre of the embankment was also lower than the ends, there being a difference of 2.55 feet between the inside edge of the embankment and the top of the waste water weir but only 1.31 feet where the water broke through. As the top of the embankment sloped downwards from the inside edge above the water to the outside edge, the front edge of the embankment at the centre was only 0.16 feet (2 inches) above the waste water weir. The embankment was formed of earth without any masonry facing or other protection, with the exception of a low dry wall of stones about 3.5 feet in depth at the top edge of the embankment and near the level of the surface of the water. This wall had apparently been washed away in several places on the southern end, and not having been repaired, the wash of the water had caused some indentations in the embankment in those places.

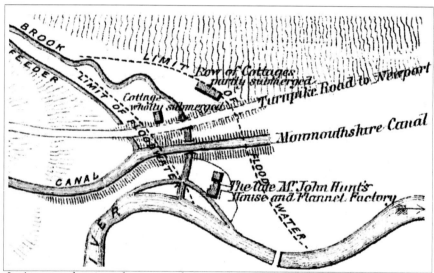

Jee's map showing the area of Cwm Carn flooded by the bursting of the reservoir. He omitted the cottage inhabited by the Davies family that lay between the aqueduct and the road bridge, its remains probably having been removed by the construction of a temporary road bridge.

There appeared to have been a bed of puddled clay in the centre of the embankment, about eight to ten feet wide at the bottom and gradually decreasing in thickness towards the top. The puddle appeared to be of very inferior quality, not of sufficiently plastic quality to hold water without risk of leakage. It may have been obtained from near the dam from excavations during construction as it was entirely different to the retentive alluvial clay from the Old Red Sandstone formation which was used from time to time for patching leaks in the embankment. Jee stated that in former years it was often

Accompanying M.ʳ George John Jee's Paper
on
THE CWM CARNE RESERVOIR

Horizontal and Vertical Scale _ 20 Feet to an Inch.

N.B. The edge of the Embankment next the Pond is numbered 1 and is drawn thus · — · — · ┼ · — · —
The outer edge of D.º is numbered 2 . . . . D.º
The Level of the Water when the Reservoir was full . . . . . . D.º
The letters A . B . &c. refer to Cross Sections

Jee's cross-section of the breach. The remaining part of the embankment lies to the left of D. The deep water culvert was not visible for an accumulation of mud. It would have been on the north side of the stream flowing through the breach and any remains have now been obscured by tipping and housing.

49

Jee's cross section of the embankment

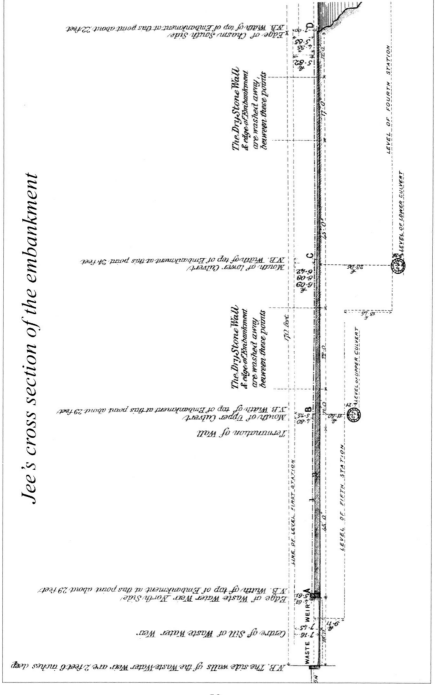

50

the practice to construct ponds and reservoirs in a very unsubstantial and imperfect manner, It was not unusual to throw an embankment of earth and stones across a valley in a suitable spot using an inferior clay which could be obtained with the least inconvenience in the immediate locality for puddle. He cited the Glyn Ponds a few miles from Cwmcarn as an exception, disused for many years but showing a marked difference in construction.

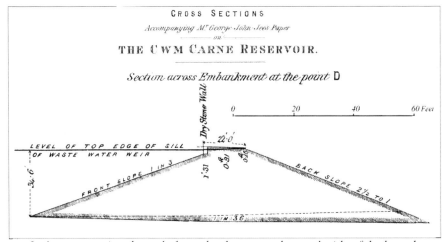

Jee's cross section through the embankment on the south side of the breach.

The breach in the embankment approximately at D on Jee's plan, some erosion having occurred to create a lesser slope than that he measured.

It was Jee's opinion that the Cwmcarn Reservoir had not been constructed on safe and approved principles, especially having regard to the great body and head of water to be retained and the mountainous nature of the district. Witnesses at the inquest stated that it had long been considered leaky and dangerous. The deep water culvert and sluice were of considerable size but did not appear to have been in working order for some time, being blocked by an old accumulation of mud. Also, it appeared that it was not accessible when the reservoir was full of water, for the top of the shaft had broken away and would have been many feet below the surface of the water without a platform to communicate with it and entirely useless as a means of providing a storm water outlet. Because of an accumulation of mud at the base of the sluice Jee was unable to measure it, but it was stated to be three feet in diameter at the inquest. The depth from the surface to the centre of the outlet was 32 feet and assuming a length of 141 feet the theoretical discharge from such an opening was nearly 320 cubic feet per second, but allowing for loss of head due to velocity of entry and friction for water passing through a culvert 141 feet long the calculated discharge would be about 184.6 cubic feet per second, or 69,000 gallons per minute. No remains of the deep water sluice, which was near the centre of the breach but would now be on the north side of the stream, are visible today because of housing development and tipping.

About 89 feet north of the waste water weir was a circular stone culvert 2.58 feet in diameter and about 132 long carried through the embankment at a depth of 23.95 feet, point C on Jee's plan. The theoretical discharge from such an opening was 205 cubic feet per second but the actual discharge would not have been more than about 115 cubic feet per second, or 43.245 gallons per minute.

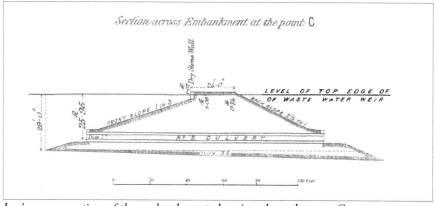

Jee's cross section of the embankment showing the culvert at C.

The remains of an almost buried culvert C on the outer face of the embankment.

The entrance to the culvert at C Jee's cross section on the inner face of the embankment. The mouth of the culvert is not visible. Stone walls protect the embankment around the mouth of the culvert, as does the lower flanking wall to the left.

There was a third culvert and sluice situated at a depth of 11.5 feet below the depth of the waste water weir, point B on Jee's plan. This had a diameter of 2.42 feet and a theoretical discharge of 124.43 cubic feet per second but an actual discharge of 78.4 cubic feet per second.

The outlet in the embankment of the culvert at B on Jee's cross section. Jee did not show a high retaining wall.

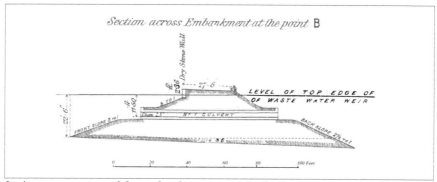

Jee's cross section of the embankment at B, the highest culvert.

The overflow or waste water weir in the south-western corner of the reservoir, point A on Jee's plan, was about 18 feet in width and 2.5 feet in depth from the level of the surface of the embankment at that point to the top of the sill. Jee said it was paved with masonry and appeared to have been properly constructed. The theoretical discharge from this outlet with a head of one foot, the depth of water when the water was overflowing the centre of the embankment, was nearly 24,000 gallons per minute, but the actual discharge after making the necessary allowances for loss of head and friction would not exceed 18,215 gallons per minute, or 43.73 cubic feet per second. Thus the two working sluices and the overflow weir could only cope with an inflow of 240 cubic feet per second. Very little stonework of the overflow remains to be seen. This area of the embankment was close to the GWR line to Cwmcarn Colliery, constructed in 1912 and now used as the access road to the 'Cwmcarn Scenic Forest Drive'. It has suffered from some wear and erosion as it is close to a much used footpath between the playing fields in front of Feeder Row and the Forest Drive which provided easy access to the dam.

The position of the edge of the waste water weir at A on Jee's cross section of the embankment is marked by this line of stones, all that remains.

On Wednesday 14th July 1875 it started raining early in the day and became increasingly heavier until Thursday morning when it slowly ceased. Jee based his calculations on rainfall of 3.74 inches measured on a gauge at Cardiff, there being no nearer one, although newspapers at the time recorded 5.21 inches of rain in 24 hours. The reservoir was fed by two streams, Nant Carn and the smaller Nant Gappy. The water of Nant Gappy passed under a road, the culvert being quite full and the water partly flowing over the road on the morning following the accident, and the bridge over Nant Carn was full up to the springing of the arch, or within 18 inches of the top early on the same day. The mean velocity of the water would have been 8 mph, being equivalent to a central surface velocity of 9.52 mph. Assuming that the quantity of water flowing into the reservoir from those sources on the evening of the flood was the same as during the following morning, the quantity of water discharged into the reservoir from Nant Gappy was equal to 112.85 cubic feet per second and from Nant Carn 422.28 cubic feet per second, making the total inflow

535.13 cubic feet per second. To this Jee added the rainfall falling on the reservoir, 3.74 inches from data collected at Cardiff over an 18 hour period, which would be equivalent to an inflow of 1.39 cubic feet per second. He estimated the area of the hillsides adjacent to the reservoir as 80 acres and calculated the allowance for surface drainage as 24.88 cubic feet per second. Thus the total inflow that day by Jee best estimates was 560 cubic feet per second based on a rainfall of 3.74 inches. He considered the sluices originally provided were barely sufficient to cope with that day's rainfall if the largest had been in working order, and it was imprudent to have allowed the reservoir to become full. The waste water weir should have been lowered by a four feet and the principal deep water sluice repaired. However, the rainfall was probably 30% higher than the figure Jee had used in his calculations, and a deeper overflow and the three sluices, if all were available, would have still been overwhelmed.

No structure of this kind could be considered safe without constant attention and supervision. Witnesses giving evidence at the inquest referred to old leaks which had been repaired by driving down puddle at the centre of the embankment, and which had been going on as long as anyone could remember. That this embankment had been causing problems for the Monmouthshire Canal Company from the time it was built is revealed by reference to leaks in the minutes of the company dated 18th October 1802 and 16th January 1805. It could be safely inferred that when a leak had been established in an embankment of this kind no puddling or repatching could affect any permanent cure, for, when a channel had been established the water under great pressure would continue to force its way through, and if stopped in one place would soon force its way out in another. The lower centre of this the Carn Reservoir embankment, as on the embankment Holmfirth, was evidence of a slump of materials due to leaks washing away material from its core.

The most likely prime cause of the failure of the dam must be water leakage through the dam from the earliest days of its construction which caused a slump in the embankment, and the neglect of essential maintenance which led to the reservoir being overwhelmed on a day of unparalleled rainfall. Warning signs were overlooked, as water was seen to be running over the top of the embankment on previous occasions during heavy floods. Jee's final conclusion is just as relevant today as when it was first made: *"It may be safely inferred that no structure of its kind, however long it may have been in existence, could be considered safe without constant attention and supervision."*

Jee's paper was read at a meeting of the South Wales Institute of Engineers but time did not allow for an immediate discussion of his findings. This took place at their following meeting by which time Jee was able to correct the rainfall figure that he had used, having seen much higher figures. On the day of the inundation there had been a fall of 5.33 inches at Newport Waterworks. At Newport it had been 5.2 inches, at Tintern 5.31 inches and 4.80 inches at Cardiff. He had used a figure of 3.74 inches at Cardiff from the *Hereford Times*, but these later figures were from *Symon's British Rainfall*. In fact, the rainfall recorded at Newport Waterworks a little to the south of Twmbarlwm would have been close to the actual rainfall over the catchment area of the Cwmcarn Reservoir. Thus all figures Jee calculated for inflowing water would have to be increased by about 30%. It was pointed out that although the situation was very convenient for the construction of a reservoir, the danger was that the streams were allowed to flow directly into it and not conducted in from a side channel controlled by sluices. It was also doubtful that there was a core of good puddle. This Jee agreed with, but he had not seen a single instance of conducting streams into reservoirs by side channels in the district, as it was more expensive. If the storm water weirs were of sufficient capacity it was unnecessary. Mr Dyne-Steel, a well-known engineer who had worked for the Blaenavon Iron and Steel Company and would have been in charge of their extensive waterworks management, said he had looked into the failure of the dam soon after it had occurred and concluded that it was due to the want of a sufficient waste water weir and sluices. The slopes were of the proper angle and the bank fairly well constructed. Indeed, for the date of its construction it was well built. He could only make the pressure on the embankment 1,610 tons as he calculated the mean depth from the sections to be 17 feet, not 24 feet. He calculated the capacity of the waste water weir as little more than half that of Jee because it was only 9.5 feet wide at its narrowest point. If water was allowed to run over an unprotected earth embankment it would inevitably cut through it in a very few hours. He said: "*The strange apathy and ignorance of the people who saw water running over the embankment, without raising an alarm, was something wonderful.*"

7

Aftermath and Reconstruction

People visited Cwmcarn for many weeks after the disaster to see for themselves the destruction caused by the flood water. Many memorial cards were printed for sale to the thousands of visitors to the scene of the disaster. Even later, in August, it was reported that copies of a photograph rescued from the debris at Cwmcarn was being sold in good numbers as of John Hunt but was actually of a family friend who was still alive. Lewis Morgan, landlord of the *Castle Inn* in Pontywaun, was charged with keeping open his house during prohibited hours on Sunday to serve many of the thousands of people that visited on the day of the funerals. Police Inspector Sheppard said Morgan was trying to make a 'harvest' out of the occasion. Many of his customers lived over three miles away and could be considered as bona fide travellers, but some were local men taking opportunity to have a Sunday drinking session. Morgan was fined £1 plus costs. Twelve men were summoned for drinking out of permitted hours: David Absolom, Henry Benjamin, Charles Parry, Henry Parry, William Jenkins, George Edmunds, David Thomas, Evan Jones alias Ebenezer Lewis, and Alfred Watkins pleaded guilty and were fined 5s. each. Two men, John Hannaford and John Rosser, did not appear and the charge against Oliver Jones was dismissed as he probably lived three miles away.

Amongst these visitors were a group of six ladies from Newport who came on Monday, 19th July, having hired a waggonette from George Pollard, omnibus proprietor of Queen Street, Newport, and driven by his sixteen year old son. They left Newport at 3.30 p.m. and on arriving at Crosskeys the horse was fed in the shafts. They left on the return journey at 7.30 p.m. and were descending Stow Hill about 8.45 p.m. when the horse went off at full pace from a trot. Pollard tried to brake with one hand and pull the horse up with the other. Near the top of Charles Street, Mrs Chapman, a shoemaker's wife and a robust and heavy woman, panicked and tried to jump out of the waggonette but fell and broke her neck.

On Sunday the parents and other family members of William Bowen left Tir Phil, New Tredegar, where his father was a collier, to visit their son who lived in one of the houses which had been flooded. Arrangements were being made to billet the family when Mrs Mary Bowen intervened following an argument between her son George Bowen and her son-in-law John Fry over sleeping arrangements and collapsed and died. Rumours spread that she had been struck in the course of the argument. The inquest was held at the *Castle Inn* on Thursday 22nd July 1875. Jane Beer of New Tredegar denied that the deceased

In Affectionate Remembrance of

The late Inhabitants of the Flannel Factory near Abercarne, viz :—

MARY ANN HUNT,	aged 52 Years,		ELIZABETH HUNT	aged 19 Years,
JOHN HUNT, jun.	„ 23 „		MARY JONES,	„ 15 „
JAMES HUNT,	„ 11 „		ELIZABETH WICKS,	„ 17 „
LETITIA HUNT,	„ 21 „		GEORGE KLIEN,	„ 15 „

And JOHN HUNT, aged 47, owner of the Factory,

(Who died on Sunday morning following from the injuries received during the flood)

ALSO OF

HOWELL DAVIES, aged 60, MARGARET DAVIES, aged 38, & JOHN DAVIES aged 34 Years,

Who met with their deaths by the bursting of Cwm Carne Reservoir on Wednesday night, July 14th, 1875.

"The Rains descended and the Floods came and beat upon that House, and it fell, and great was the fall of it."—Matt., vii., 27.

(J. H. Taylor, Machine Printer, Risca.

IN MEMORY OF
JOHN HUNT, AGED 49,

(Who died July 18th, 1875),

MARY ANN HUNT, AGED 52, (Wife,)

JOHN HUNT, 22,   **LETITIA M. HUNT, 21,**
SON.   DAUGHTER.

ELIZABETH J. HUNT, 19   **JAMES HUNT, 10,**
DAUGHTER.   SON.

Who met with their deaths by the bursting of Cwm Carne Reservoir, on Wednesday night, July 14th, 1875, and were interred at the Independent Chapel, Pontywain.

"The Rains descended and the Floods came and beat upon that House, and it fell, and great was the fall of it."—Matt. vii., 27 v.

Two of the many memorial cards printed for sale as mementos to the many thousands of visitors who flocked to the scene of the disaster at Cwmcarn.

had been stuck and said Mary Bowen was holding John Fry on the chair when she collapsed and died. She was subject to fits when excited. It was stated that the men had had beer but were not tipsy. The coroner's jury recorded a verdict that Mary Bowen died in a fit whilst under excitement, and there was no evidence to show that any violence had been used.[20]

60

The Castle Inn lies opposite the canal and near the head of the remaining section of the Crumlin arm of the canal.

Although the General Committee of the Monmouthshire Railway and Canal Company met on the 15th July 1875, the day after the disaster, the minutes of that meeting contain no record of the event or any discussion that took place. At a meeting of the Sub-Committee the following week, the Company's engineer reported on the failure of the Cwm Carn Reservoir, sundry slips and damage to the railways, but no details were recorded. Edwin Jones and W. Roberts, boat owners and carriers on the Crumlin line of the canal asked to be compensated for their loss of business through the canal having been destroyed. The General Committee of the Company met on the 16th August and resolved: *"That they be informed that the Committee very much regret their being sufferers by the disaster which occurred to the Canal on the night of the 14th of July, but as this company cannot be held responsible for the effect of such an extraordinary fall of rain the claim to compensation cannot be entertained; also that it is the intention of this Company to re-instate the Canal with all reasonable dispatch."*

Mr T. Llewellin of Newport, solicitor to the Abercarn Turnpike Trust, hurried to the scene of the disaster and instructed Mr Sheppard, the trust's surveyor, to have a temporary footbridge erected so that ordinary traffic could be resumed. This was carried out by Thomas Williams, builder, of Cwmdows.[21] Orders were given for the construction of a new bridge as soon as possible. A large deputation representing the interests of the Abercarn Turnpike Trust was invited to attend a meeting of the Sub-Committee of the canal company on the 19th August 1875. Their spokesman, Mr C. Parkes, stated that they considered the company were liable for the destruction of the road at Carn Mill, and did not wish to take hostile proceedings if the Company would assist by a liberal contribution in restoring the road, the cost of which would be £550. An improvement could be effected by a gift of some land at a cost of £630. It was decided that the matter would be put before the board, and Mr Warren was asked to supply Mr Harrison of the canal company with further information. The General Committee authorised Mr Harrison to negotiate with Mr T.M. Llewellin, clerk to the trustees, for a settlement of their claim.

Nearly two months later there was little to show from any negotiations. Parish meetings had been held in Risca and Mynyddislwyn on the subject of damage to the road and those meetings had decided to make no claim on the company if the company would surrender a certain piece of waste land between the road and the canal for the purpose of the road improvement. This would allow the construction of a combined aqueduct and road bridge over Nant Carn, the structure which survives to this day. Llewellin attended a sub-Committee meeting to inform them of these views and it was agreed that he and Mr Gustard should meet and settle an agreement and assignment of this land.

Factory Trip and the new combined aqueduct and viaduct, about 1890. Two ladies in their flannel shawls pose for the photographer, one having put down her can probably used for the purchase of lamp oil.

The whereabouts of the minutes of the Abercarn Turnpike Trust, abolished in 1879, and its successor the Mynyddislwyn District Highway District are unknown and almost certainly have not survived. Between Newbridge and Newport the Abercarn Turnpike Trust was responsible for the road as far as St.

The 1880 Mynyddislwyn Highway District boundary marker at the centre of the rebuilt road bridge and aqueduct.

Mary's Church in Risca and the Newport Turnpike Trust for the remainder. The boundary between the Highway Districts that replaced the turnpike trusts at the end of 1879 was the middle of the Carn Bridge. The committee of the Newport Turnpike Trust met on Wednesday 11th August 1875, two weeks after the disaster. Present at the meeting were their clerk, surveyor and three trustees, Samuel Homfray and the Reverends Hawkins and Jenkins. Although this road must have been damaged and needed stones and mud cleared, homes in Risca having been flooded with water waist deep, the minutes state only that the surveyor's report on the damage to their road by the late heavy flood was read. Details of the damage or cost of repairs are not recorded, nor did trustees apparently visit to inspect the damage.[22] This appears surprising considering the damage that must have been caused by the flood water of 1875. Yet a year later, at a meeting held on 11th October 1876 to consider road damage caused by an overflowing stream at Risca, the surveyor reported that repair of the damage done by the flood would cost £30 and that some steps should be taken to get the brook by the side of the road cleaned out. The clerk was directed to ascertain who was responsible for this work. Furthermore, when a Mr Dixon wrote to the Newport Turnpike Trust complaining of the state of the road at Pontymister, a sub-Committee was formed to inspect the road. Their surveyor, clerk and three trustees attended on 10th February 1877 and reported that the road was in a dirty state, buildings were being built too close to the road and that a drain was needed on its eastern side.

Both the Rev. J. W. Winspear and Lt. Col. Heyworth of Waunfawr House, Crosskeys, suggested that a public meeting should be held towards alleviating the suffering of those rendered homeless and penniless, and the erection of a memorial stone over the grave of the Hunt family. A public meeting was held on Monday 19th July 1875 and the *Cwmcarn and Risca Inundation and*

Memorial Fund was set up with the object of erecting a much needed cottage hospital in the area as a memorial to those who died. Lt. Col. Lawrence Heyworth J.P. was elected chairman, Thomas Moses of Risca became secretary, and S. Vernon, manager of the Newport branch of the West of England Bank, was appointed honorary treasurer. Other elected members of the Committee were the Rev. Hugh Williams of Risca, the Rev. J.W. Winspear, Wesleyan minister of Risca, the Rev. Peter H. Davies, minister of Trinity Independent Chapel, Pontywaun, David Morris of Danygraig and proprietor of the chemical works, Risca, G.H. Banks of the Pontymister tinplate works, Thomas Vaughan of Risca and manager of the Pontymister tinplate works, Jonathan Piggford, George Jones, William James and William Hartshorn, all of Risca, and Henry Jones and Joseph Small of Pontywaun. It was even suggested that surplus money in the Risca Colliery Disaster widows and orphans fund should be used to relieve distress, as a handsome sum was still in the hands of bankers that would never be used for the purpose for which it was raised.[23]

Colonel Heyworth wrote to the canal company soliciting a subscription in aid of a proposed cottage hospital at Abercarn, which was read out at a meeting of the General Committee chaired by Lord Tredegar held on the 21st December 1875. They decided that the funds of the company were not available for that purpose. None of the money raised by the *Cwmcarne and Risca Inundation and Memorial Fund* was ever put towards its charitable aims. The memorial on the grave of the Hunt family in Trinity churchyard was erected by their relatives. None of the other victims buried in Trinity churchyard have memorials erected over their graves. No cottage hospital was ever built, but one would have been of great use in treating the many victims of colliery accidents. A few years ago, by permission of the Charity Commission, that fund was applied to community projects in the area.

Colonel Heyworth's claim, amounting to £117 10s., against the canal company for damage to his land and property was also refused, as were those of many other claimants. At a meeting of the General Committee, comprising John Lawrence (in the chair), the Rt. Hon. Lord Tredegar, William Evans, Thomas Brown, Thomas Gratrex, George Cave, Charles W. Savage, Edward J. Phillips and George Miller, it was resolved: *"That Mr. H. Brain be informed that the funds of the Company are not available for the purpose of assisting him to rebuild the Flannel Factory which was destroyed by flood water from Cwm Carn Reservoir on July 14th last."* H. Brain was the brother of John Hunt's wife Elizabeth.

In 1881 the only survivor of the disaster James Foley, then 21 years of age, was boarding with Jonah Mayler, a flannel weaver, in Pembroke Dock and working as a weaver. In 1891 he was working as a collier in Ystradyfodwg in the Rhondda Valley.

The flannel factory was eventually rebuilt and operated by Mrs Catherine Johns in 1881, and later by her son William Johns until it closed, being listed for the last time in *Kelly's Directory* for 1920. Many similar flannel mills closed in the latter half of the nineteenth century as the growing railway network allowed the importation of cheaper machine-made cloth from the North of England and the sale of ready-made clothes. Cwmcarn flannel mill was marked as disused on the 1920 edition of the Ordnance Survey map. 1:2,500 scale. Today, one can still walk under the arch of the new aqueduct but nothing remains of the flannel mill.

The rebuilt flannel factory at Cwmcarn, early 1900s. The tentering frame for drying the flannel can be seen on the left. The line of the leat providing water power from the river is just visible mid left.

The remains of the embankment of the canal company's reservoir are now a Scheduled Ancient Monument, reference number MM259/AM41.

Appendix

Prior to the breach of the Cwmcarn Reservoir there had been two major disasters as a result of the failure of earth dams. The first of these was the Holmfirth flood on 5th February 1852. Holmfirth is familiar to viewers of BBC television as the location for the programme *The Last of the Summer Wine*. It was a mill town situated on the banks of the River Holme, and in 1837 the mill owners obtained an Act of Parliament to regulate the supply of water to their mills. These were worked by water power and suffered in summer droughts. At Bilberry Mill, about three miles above Holmfirth, a reservoir was formed in 1840 by an embankment 340 feet long and 98 feet high across the Digley Brook at the top of a narrow glen, having a water surface of 15˙20 acres. The dam had a core of puddle 16 feet thick at the base and 8 feet thick at the top, with a wall of rubble on either side.[24] Heavy rains had prevailed for some days in the district for some days but before nightfall on 4th February the clouds had moved away. The centre of the embankment was lower than the overflow, and as winds and rising water beating upon the embankment began to wash parts of it away and create deep fissures. It gave way on the 5th February just before 1 a.m. as the residents of the villages in the valley and Holmfirth below slept in their beds, unaware of the impending disaster. Witnesses described a rumble like thunder and a bank of water rushing down the valley. The water took about 20 minutes to reach Holmfirth and the reservoir was emptied in less than 30 minutes. 81 men, women and children died. 244 mills, houses, shops, public houses, churches and other properties were destroyed or seriously damaged and over 7,000 adults and children were thrown out of work.

The verdict of the jury at the Coroner's inquest on one of the victims of the Holmfirth flood was: *"We find that Eliza Marsden came to her death by drowning, caused by the bursting of the Bilberry reservoir. We also find that the Bilberry reservoir was defective in its original construction, and that the Commissioners, engineers and overlooker were greatly culpable in not seeing to the proper regulation of the works; and we also find, that the Commissioners in permitting the Bilberry reservoir to remain for several years in a dangerous state, with the full knowledge thereof, and not lowering the waste pit, have been guilty of gross and culpable negligence: and we regret that the reservoir, being under the management of a corporation, prevents us from bringing in a verdict of manslaughter, as we are convinced that the gross and culpable negligence of the Commissioners would have subjected them to such a verdict, had they been in the position of a private individual or firm. We also hope that the Legislature will take into its most serious consideration, the propriety of making provision for the protection of the lives and properties of her Majesty's subjects, exposed to dangers from reservoirs, placed by corporations in situations similar to those under the charge of the Holme Reservoir Commissioners.*" Despite the jury finding there had been culpable negligence on the part of the officials and Commissioners controlling the reservoir, no action was taken. Parliament did not introduce any new legislation to control earth dams. The Coroner's jury were very forthright in their verdict and sensibly called for legislature in order to eliminate future accidents of this nature. The circumstances of the flood bear great similarity to the disaster at Cwmcarn which might have been avoided had steps been taken to inspect earth dams and their maintenance.

The Rev. Josiah Bateman, vicar of Huddersfield, said of the Holmfirth flood:[25] "*And though it may not be easy to affix the blame on any one individual, yet it is agreed on all hands that the work was badly done. An embankment, to be safe, must be sound; and here there was no soundness. A spring broke out at its base whilst it was being made, which defied the ill-judged remedies adopted, and gradually sapped its very foundations. Several serious leakages also appeared in various parts of the structure itself, and added to its insecurity. Moreover it began to settle directly it was finished and continued to do so till the top of it in the middle part became eventually lower than the top of the waste pit. This caused the final catastrophe; for the provision made for carrying harmlessly away the surplus water was rendered useless. Instead of running down the waste pit, it ran first over the embankment, sunk down through the hollows caused by the settling into its very substance, saturated the earth of which it was constructed, finally boiled up, and burst away. The results of the legal enquiry satisfactorily establishes the fact, that all the terrible results which have filled the valley with mourning might have been averted, had timely precautions been taken, and the waste pit lowered. The very Commissioners themselves seem to have been aware of this; and so long ago as 1846, they passed an order to that effect. But the order was not executed, and the non-execution of it was not reported. Some parties were unwilling to give offence to others, some had no time to spare, some would not work without pay, and in short no one seemed to consider himself in any wise responsible; and whilst the water was in constant use, the embankment was left to take care of itself.*" Much of this was said about the Cwm Carn Reservoir and its embankment at the inquest into the cause of death of the twelve victims of that disaster.

The second major failure of an earthen dam was in 1859 near Sheffield. By an Act of Parliament obtained in 1853 the Sheffield Waterworks Company were empowered to build a reservoir in the headwaters of the Loxley River, about eight miles above Sheffield.[26] Their consulting engineer was the nephew of the man who built the Bilberry Mill reservoir near Holmfirth, little more than nine miles away. The dam was 418 yards long, 95 feet high in the centre, 500 feet wide at the base and 12 feet wide at the top, with a central core of 'puddle'. The reservoir so formed covered about 78 acres. Construction of the Dale Dyke dam began on 1st January 1859. During construction hidden springs had caused difficulties in forming the water-tight core. There had been talk of minor subsidence and that insufficient time had been allowed for the settling of materials. By 11th March 1864 the dam was largely completed and the reservoir had been allowed to fill. Water was within a few feet of the top when the engineer's attention was drawn to a small crack in the face of the dam. Just before midnight a high wind whipped up the water and the dam was breached. Within seconds the dam collapsed and a wall of water swept down the valley towards Sheffield, destroying everything in its path. Occupants of buildings in its path were drowned or killed by the debris hurled along by the force of the water. Over 250 people died that night.

After the inquest the jury's verdict on one of the victims was: *"We find that Thomas Elston came to his death by drowning in the inundation caused by the bursting of the Bradfield reservoir, on the morning of the 12th of March instant; that, in our opinion, there has not been that engineering skill and that attention to the construction of the works, which their magnitude and importance demanded; that, in our opinion, the Legislature ought to take such action as will result in a Government inspection of all works of this character, and that such inspection should be frequent, sufficient and regular; that we cannot separate without expressing our deep regret at the fearful loss of life which has occurred from the disruption of the Bradfield reservoir."* Most people accepted that errors had been made in the construction of the Dale Dyke dam but no specific reason for its collapse was proved. The coroner was said to have been too outspoken, and the jury's verdict was criticised as vague as nobody in particular was condemned.

No legislation was introduced for the inspection of reservoirs after this disaster. The Cwmcarn reservoir and its embankment was a disaster waiting to happen.

References

Abbreviations used:

GwRO	Gwent Record Office
NLW	National Library of Wales
TNA	The National Archives, Kew

The accounts of the disaster, funerals, inquest and connected events are from the *Monmouthshire Merlin* and *the Star of Gwent* newspapers of 17th, 24th and 31st July and unless otherwise indicated, available on microfilm at Newport Reference Library.

1. *The Canals of South Wales and the Border,* by Charles Hadfield. University of Wales Press, Cardiff, 1960, p. 127.
2. GwRO, Misc. MSS, 1388.
3. TNA, RAIL 500/5. Monmouthshire Canal Company Committee minute book, 1792-1812.
4. TNA, RAIL 500/1. Monmouthshire Canal Company General Assembly Proceedings, 1794-1843.
5. NLW, Tredegar 50/133.
6. NLW, Tredegar 43/117.
7. *Hall's Tramroad,* by Foster Frowen. Published in 'Archive, The Quarterly Journal for British Industrial and Transport History' in 5 parts, Issue **55**, p.27-37; **56**, 31-54; **59**, 25-39; **60**, 17-39; **66**, p. 3-33.
8. TNA, RAIL 500/7. Monmouthshire Canal Company Committee minute book, 1831-1849.
9. *"Glimpses of West Gwent"* by Rex Pugh, R.H. Johns, Newport, 1934.
10. GwRO, D32.333.
11. *Monmouthshire Merlin,* May 21st and 28th 1852. *London Gazette,* 25th September 1855 and 8th December 1857.
12. GwRO, Trevethin parish registers, D.Pa 13.17, entry no. 482.
13. *Western Mail*, 16th July 1875.
14. *My Life's History: The Autobiography of Rev. Thomas Lewis.* Edited by W. Edwards, D.D., published by Miss C.E. Lewis, Newport, 1902, p. 137.
15. *Western Mail*, 17th July 1875.
16. *Western Mail*, 24th July 1875.
17. TNA, RAIL 500/18. Monmouthshire Railway and Canal Company minute book, 1873-1875.
25. *Victorian Engineering,* L.T.C. Rolt,
19. Proceedings, South Wales Institute of Engineers, No. 3, Vol. 10,

1877, p.136. and Discussion, p. 217.

20.	*Western Mail*, 23rd July 1875.
21.	*Western Mail*, 22nd July 1875.
22.	GwRO, Microfilm 237.
23.	*Western Mail*, 4th August 1875.
24.	*Complete Account of the Holmfirth Flood*, Eli Collins & Co. Ltd., Holmfirth, 1910.
25.	*The Holmfirth Flood – A Narrative*, by Rev. Josiah Bateman, published by I. Seeley, London, 1852.
26.	*The Collapse of the Dale Dyke Dam 1864,* by Geoffrey Amey, Casell, London, 1974.